Band 22 · März 1986

# 3

# isotopenpraxis

AKADEMIE-VERLAG · BERLIN

ISSN 0021-1915 Isotopenpraxis, Berlin 22 (1986) 3, 73—108

## An unsere Autoren

Die Zeitschrift veröffentlicht Beiträge zu folgenden Themen: Kontrolle und Analyse radioaktiver und stabiler Isotope, Anwendung radioaktiver und stabiler Isotope in allen Gebieten der Forschung und Technik, der Industrie, Landwirtschaft und Medizin, Herstellung radioaktiver und stabiler Isotopen sowie von markierten Verbindungen, kernelektronische Geräte für die Isotopenanwendung (Dickenmesser, Dichtemesser u. ä.) Detektoren für isotopentechnische Arbeiten und den Strahlenschutz in Isotopenlaboratorien, Meßtechnik und Strahlenschutz, Dosimetrie in Isotopenlaboratorien, kerntechnische Geräte für die Isotopenanwendung (Beton, Bestrahlungsanlagen u. ä.), Strahlenchemie, Radiochemie, ökonomische und perspektivische Fragen, Fragen der Weiterbildung.

### *1. Aufnahme von Manuskripten*

Zur Veröffentlichung angenommen werden nur druckreife, noch nicht anderweitig publizierte Beiträge in deutscher, englischer oder russischer Sprache zum Themenbereich der Zeitschrift und zwar

- Originalarbeiten
- Kurze Mitteilungen
- Aus der Praxis
- Zusammenfassende Berichte
- Tagungsberichte
- Informationen.

Sind der Beitrag oder Teile daraus bereits veröffentlicht oder ist eine anderweitige Veröffentlichung vorgesehen, so muß dies der Redaktion mit genauer Angabe der Zweitveröffentlichung mitgeteilt werden. Mit der Annahme des Manuskriptes behält sich der Verlag das Recht der Vervielfältigung, Verbreitung und Übersetzung innerhalb der gesetzlichen Schutzfrist vor. Auszugsweiser Nachdruck oder photomechanische u. a. Wiedergabe sind an eine schriftliche Genehmigung des Verlages gebunden. Manuskripte, Zeichnungen und Photos von veröffentlichten Arbeiten werden nur auf den besonderen und bis zum Zeitpunkt der Veröffentlichung mitgeteilten Wunsch des Autors zurückgesandt.

Bei der unverlangten Einreichung von Manuskripten wird empfohlen, eine Befürwortung eines der Herausgeber oder Redaktionsausschußmitglieder beizufügen.

Manuskripte sind einzureichen bei der Redaktion, 1086 Berlin Leipziger Str. 3–4

### *2. Form der Manuskripte*

Die Autoren werden gebeten:

2 Manuskripte, eines davon in Original-Maschinenschrift, einseitig und mit 2zeiligem Abstand einzureichen; gut leserliche Photokopien sind zulässig; auf dem 1. Blatt (unter Weglassung der Bezeichnung)

den Titel der Arbeit
Vorname-Initialen und Namen sämtlicher Autoren
gegebenenfalls die Institution, in der sie tätig sind
eine kurze Zusammenfassung der Arbeit von höchstens 15 Zeilen
Keywords in alphabetischer Reihenfolge und
als Fußnote die Anschrift der Institution anzugeben;

wenn möglich, Zusammenfassung, Titel und Autoren *englisch* (bei deutschem oder russischen Manuskript) bzw. *deutsch* (bei englischem Manuskript) 2fach beizufügen;

Literaturhinweise (fortlaufend numeriert), Fußnoten, Tabellen, Abbildungen (in 2–3facher Veröffentlichungsgröße), Abbildungsunterschriften (getrennt von den Abbildungen und Zeichenerklärungen, Quellenangaben, falls erforderlich, und gegebenenfalls mit einem Vermerk für besondere Wünsche bezüglich der Abbildungsgröße), private und Dienstanschrift der Verfasser mit akademischen Titeln und ausgeschriebenen Vornamen auf besonderen Blättern ans Ende des Manuskriptes zu stellen;

Wünsche für den Kolumnentitel (Seitenüberschrift) bekanntzugeben;

Absätze durch Einzüge von mindestens 3 Anschlägen kenntlich zu machen; die Maßeinheiten des SI-Systems („Systeme International") zu verwenden; Korrekturen im Manuskript unmittelbar an der zu korrigierenden Stelle anzubringen;

Formeln deutlich lesbar und möglichst groß einzutragen;

Farbabbildungen und nachträgliche Änderungen oder Ergänzungen in Abbildungen zu vermeiden;

hervorzuhebende Textstellen zu unterstreichen, und zwar halbfetten Satz mit einer geraden schwarzen Linie, kursiven Satz (Schrägschrift) mit einer blauen Wellenlinie (alle Eigennamen). Sperrungen gestrichelt;

besondere Buchstaben durch farbige Unterstreichung zu kennzeichnen, und zwar Formelzeichen in

Griechisch-Kursiv ($\alpha$, $\beta$, ...) rot,
Kursiv ($A$, $B$, ...) blau
halbfetter Kursiv (**A**, **B**, ...) 2 Linien blau (Vektoren),
halbfetter Grotesk (**A**, **B**, ...) 2 Linien violett (Tensoren);

bei Literaturangaben auf Vollständigkeit zu achten. In nachstehender Reihenfolge müssen angegeben sein:

a) bei *Zeitschriften*: Vorname-Initialen und Namen sämtlicher Autoren, Titel der Zeitschrift (nach „Periodica Chimica" oder möglichst wenig gekürzt), Bandnummer (unterstrichen), Jahreszahl (in Klammern), Heftnummer (nur bei Zeitschriften, die jedes Heft mit S. 1 beginnen) sowie die Seitenzahl, auf der die zitierte Arbeit beginnt.

b) bei *Büchern*: Initialen der Vornamen und Name des Autors (Herausgeber werden durch den Zusatz „ed." gekennzeichnet), voll ausgeschriebener Titel des Buches in der Originalsprache, evtl. Bandnummer, voll ausgeschriebener Name des Verlages, Nummer der Auflage (außer der ersten), Erscheinungsort(e), Erscheinungsjahr (hier nicht in Klammern setzen!) und evtl. Kapitelnummer, Seitenzahl u. ä.

c) bei *Sammelbänden, Tagungsberichten*: Es soll zusätzlich der Titel des zitierten Beitrags angegeben werden. Die Seitenzahl muß unbedingt angegeben werden. Aus dem Zitat muß hervorgehen, ob es sich um einen Tagungssammelband oder um einen noch nicht veröffentlichten Vortrag handelt. Zwischen Tagungsort und Erscheinungsjahr des entsprechenden Tagungsberichts ist zu unterscheiden.

### *3. Korrekturen*

Der Autor erhält zwei Umbruchabzüge, von denen einer möglichst schnell korrigiert an die Redaktion zurückgereicht werden soll. Auf Wunsch des Autors kann die Korrektur auch von der Redaktion gelesen werden.

Durch nachträgliche Manuskriptänderungen verursachte Korrekturkosten gehen, sofern sie 10% der Satzkosten übersteigen, zu Lasten des Autors.

Der Autor erhält 50 Sonderdrucke kostenlos. Weitere Sonderdrucke können bis zur Rückgabe der Korrekturabzüge auf Kosten des Autors bei der Redaktion bestellt werden.

## Bezugsmöglichkeiten

Bestellungen sind zu richten

- **in der DDR** an den Postzeitungsvertrieb unter Angabe der Kundennummer des Bestellenden oder an den AKADEMIE-VERLAG BERLIN, DDR-1086 Berlin, Leipziger Straße 3–4, PF-Nr. 1233;
- **im sozialistischen Ausland** an eine Buchhandlung für fremdsprachige Literatur oder an den zuständigen Postzeitungsvertrieb;
- **in der BRD und Berlin (West)** an eine Buchhandlung oder an die Auslieferungsstelle KUNST UND WISSEN, Erich Bieber OHG, Wilhelmstraße 4–6, D-7000 Stuttgart 1;
- **in den übrigen westeuropäischen Ländern** an eine Buchhandlung oder an die Auslieferungsstelle KUNST UND WISSEN, Erich Bieber GmbH, General-Wille-Str. 4, CH-8002 Zürich
- **im übrigen Ausland** an den Internationalen Buch- und Zeitschriftenhandel; den Buchexport Volkseigener Außenhandelsbetrieb der Deutschen Demokratischen Republik, DDR-7010 Leipzig, Postfach 160; oder an den AKADEMIE-VERLAG BERLIN, DDR-1086 Berlin. Leipziger Straße 3–4.

Zeitschrift „Isotopenpraxis"
Herausgeber: Zentralinstitut für Isotopen- und Strahlenforschung der Akademie der Wissenschaften der DDR.
Verlag: Akademie-Verlag Berlin, DDR-1086 Berlin, Leipziger Straße 3–4, PF-Nr. 1233; Fernruf: 2236221 oder 2236229; Telex-Nr.: 114420;
Bank: Staatsbank der DDR, Berlin, Kto.-Nr.: 6836-26-20712.
Chefredakteur: Prof. Dr. H. Hübner, Leipzig.
Anschrift der Redaktion der Zeitschrift Isotopenpraxis: Prof. Dr. Hübner, DDR-7010 Leipzig, Karl-Marx-Platz, Hauptpostamt, PSF 370; Fernruf: 281018.
Veröffentlicht unter der Lizenznummer 1527 des Presseamtes beim Vorsitzenden des Ministerrates der Deutschen Demokratischen Republik.
Gesamtherstellung: VEB Druckerei „Thomas Müntzer", DDR-5820 Bad Langensalza.
Erscheinungsweise: Die Zeitschrift „Isotopenpraxis" erscheint monatlich. Die 12 Hefte eines Jahrganges bilden einen Band. Bezugspreis je Band 240,— DM zuzüglich Versandspesen; Preis je Heft 20,— DM. Der gültige Jahresbezugspreis für die DDR ist der Postzeitungsliste zu entnehmen.
Bestellnummer dieses Heftes: 1079/22/3.

AN (EDV) 12831
00900

Isotopenpraxis 22 (1986) 3, S. 73 – 80

# The Kinetic Deuterium Isotope Effect of Proton Transfer Reactions in Solution

*J. Sühnel*

(Sektion Chemie der Friedrich-Schiller-Universität Jena)[1])

*The theoretical models for the treatment of the kinetic isotope effect of proton-transfer reactions (tunneling, Marcus and quantum-statistical mechanical model) are described. Scope and limits of these models are discussed with respect to intermolecular interaction, free energy relations, temperature, solvent and pressure dependence, and ultrafast proton-transfer.*

*Die theoretischen Modelle zur Behandlung des kinetischen Isotopie-Effektes von Protonentransfer-Reaktionen werden beschrieben. Es werden Möglichkeiten und Grenzen dieser Modelle im Hinblick auf die intermolekulare Wechselwirkung, Freie-Enthalpie-Beziehungen, Temperatur-, Lösungsmittel- und Druckabhängigkeit, sowie auf den ultraschnellen Protonentransfer diskutiert.*

### Keywords

activation energy; deuterium; intermolecular forces; isotope effects; pressure dependence; reaction kinetics; statistical mechanics; tunnel effect

## 1. Introduction

The broad interest in the kinetic deuterium isotope effect (KIE) of proton-transfer reactions originates from at least two facts. At first the investigation of the KIE is widely used for the elucidation of reaction mechanisms and at second its occurrence is intimately related to the phenomenon of tunneling in chemical reactions [1 – 3].

The aim of this paper is to discuss recent developments concerning the theoretical models used and to interrelate the theoretical results to experimental data. It is, however, beyond the scope of this paper to give a full account of the experimental and theoretical work. It is rather intended to present a discussion on advantages and disadvantages of the different theoretical models in the light of experimental facts. In section 2 a short review of the theoretical models available is presented and in section 3 theoretical results are compared to experimental data. The discussion is exclusively confined to the primary isotope effect in direct proton-transfer. Solvent participation and secondary or solvent isotope effects are not considered.

## 2. Theoretical models

There are at present essentially three theoretical models for the treatment of proton-transfer reactions in solution: the tunneling model (TM), the *Marcus* model (MM) and the quantum-statistical mechanical model (QSM).

### 2.1. Tunneling model

As early as 1935 *R. P. Bell* put forward the idea that the tunneling phenomenon should also be relevant to chemical reactions, especially to the transfer of light atoms [2, 3]. According to the *Bell* approach (a one-dimensional treatment) the semiclassical rate constant $k_s^L$ (L = H, D) has to be multiplied by a tunnel correction factor $Q_t^L$, eq. (1).

$$k_L = Q_t^L k_s^L \tag{1}$$

[1]) Address: DDR-6900 Jena, Philosophenweg 14

In eq. (1) $k_s^L$ is any theoretical rate constant, which does not take into account the tunnel effect. Usually $k_s^L$ of the transition state theory is used. $Q_t^L$ is calculated by integration over the energy $W$ thus assuming a continuous distribution of states, Fig. 1 and eq. (2).

$$Q_t^L = (1/k_BT)\exp(-V_0^L/k_BT)\int_0^\infty G^L(W)\exp(-W/k_BT)\,dW \tag{2}$$

In eq. (2) $G^L$ is the barrier permeability, $V_0^L$ the barrier height, $k_B$ the *Boltzmann* constant and $T$ the temperature. *Bell* has shown that provided the barrier shape is represented by a symmetric truncated parabola $Q_t^L$ can be written as

$$Q_t^L = \frac{(\pi\beta/\alpha^L)}{\sin(\pi\beta/\alpha^L)} - \sum_{n=0}^{\infty} \frac{(-1)^n\,\beta\exp\{[\beta-(n+1)\,\alpha^L]V_0^L\}}{(n+1)\,\alpha^L-\beta} \tag{3}$$

($\alpha^L = 2\pi/h\nu_L$, $\beta = 1/k_BT$) [2, 3]. The barrier frequency $\nu_L$ is related to $V_0^L$ and to the barrier width $2b_L$ by eq. (3), see also fig. 1.

$$V_0^L = 2m_L\pi^2 b_L^2 \nu_L^2 \tag{4}$$

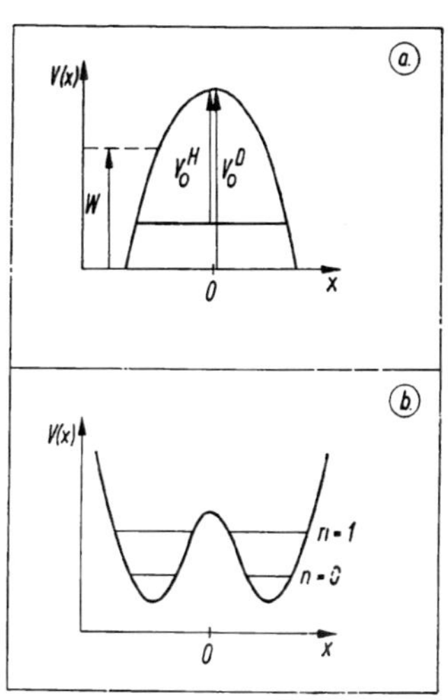


Fig. 1. Tunneling models
a – *Bell* approach; b – tunneling model for a small number of states

In eq. (4) $m_L$ is the mass of the proton or deuteron. In all papers utilizing a parabolic and not an *Eckart* barrier shape $Q_t^L$ is calculated according to eq. (3) considering, however, only the first or the first few terms of the infinite series.

Recently a new analytic approximation for the calculation of $Q_t^L$ for a parabolic barrier was presented, which provides a numerically accurate alternative even to numerical integration eq. (5)–(7) [4].

$$Q_t^L = \frac{(\pi\beta/\alpha^L)}{\sin(\pi\beta/\alpha^L)} - \frac{\beta}{\alpha^L - \beta} \exp[(\beta - \alpha^L)(V_0^L - V_2)]$$

$$(\beta \leqq \alpha^L) \tag{5}$$

$$Q_t^L = \frac{\beta}{\beta - \alpha^L} \{\exp[(\beta - \alpha^L)(V_0^L - V_2)] - 1\} \quad (\alpha^L \leqq \beta) \tag{6}$$

$$V_2 = \max(V_1, 0) \tag{7}$$

As far as unsymmetric barriers are concerned the result of the symmetric barrier carries over to an exergonic barrier with no modification. For endoergonic barriers eq. (5)–(7) with $V_1 > 0$ have to be applied.

The KIE for $k_s^L$ arises from the difference of the zero point energy for H and D vibrations in the initial and the transition state. The isotope dependence of $Q_t^L$ can be seen from eq. (5)–(7). The *Bell* approach gives the following basic results, Fig. 2. The *Arrhenius* plot is non-linear and this non-linearity is different for H and D. At low temperatures the rate constant is nearly independent of temperature though a considerable barrier of activation may exist. The KIE increases with decreasing temperature and arrives finally a plateau-like region. The temperature dependence of the isotope effect for the *Arrhenius* preexponential factors $(A_D/A_H)$ and activation energies $(E_a^D - E_a^H)$ displays maxima. The position of these maxima is dependent on the dimension of the tunnel barrier. These dimensions are usually adjusted so as to fit the experimental *Arrhenius* plot. This procedure gives in the vicinity of room temperature isotope effects for $k_L$ and the *Arrhenius* parameters larger than the semiclassical values.

The *Bell* approach is frequently applied even to intramolecular proton-transfer though in this case the assumption of a continuous distribution of states is certainly not justified and integration should be replaced by summation. In this case the rate constant may be written as

$$k_L = \sum_n p_n^L k_n^L \tag{8}$$

where $k_n^L$ stands for the tunnel frequency of the state $n$ and $p_n^L$ is the corresponding population, see Fig. 1. $k_n^L$ can be calculated according to the resonance splitting or the transmission function method [5]. A simplified procedure utilizing a triangular barrier was also proposed [6]. A systematic comparison of the TM for a small and an infinite number of states has not yet been done. The transition between the temperature dependent and the temperature independent range should, however, be more abrupt in the first case.

The TM is usually applied to the temperature dependence of the KIE. It was, however, pointed out by *Bell* that it gives also the maximum isotope effect for thermoneutral reactions predicted by the MM and the QSM, as in this case the barrier to tunneling is largest.

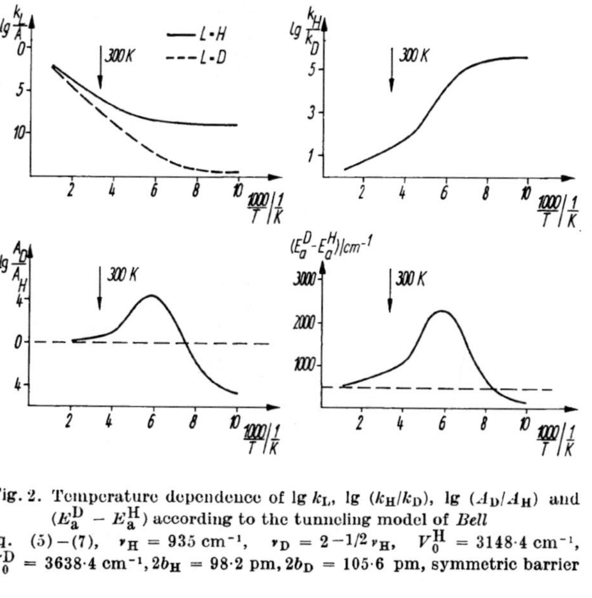


Fig. 2. Temperature dependence of lg $k_L$, lg $(k_H/k_D)$, lg $(A_D/A_H)$ and $(E_a^D - E_a^H)$ according to the tunneling model of *Bell* eq. (5)–(7), $\nu_H = 935\ cm^{-1}$, $\nu_D = 2^{-1/2}\nu_H$, $V_0^H = 3148{\cdot}4\ cm^{-1}$, $V_0^D = 3638{\cdot}4\ cm^{-1}$, $2b_H = 98{\cdot}2$ pm, $2b_D = 105{\cdot}6$ pm, symmetric barrier

## 2.2. *Marcus model*

The *Marcus* model originally designed for electron-transfer reactions and later-on also applied to proton-transfer has a completely different physical basis [7]. Whereas in the TM the activation energy stems from the stretching of the corresponding X-L bond, in the MM the main part of the activation energy originates from the reorganization of the medium and low-frequency intramolecular modes, eq. (9)–(11).

$$k_L = Z \exp[-(\Delta G^{\neq})^L/k_B T] \tag{9}$$

$$(\Delta G^{\neq})^L = w_r + \frac{\lambda_L}{4}\left[1 + \frac{\Delta G_0'}{\lambda_L}\right]^2 = w_r + \frac{(\lambda_L + \Delta G_0')^2}{4\lambda_L} \tag{10}$$

$$\Delta G_0' = w_p - w_r + \Delta G_0 \tag{11}$$

The free energy of activation $(\Delta G^{\neq})^L$ depends on the free energy of reaction $\Delta G_0$, on the so-called work terms $w_r$ and $w_p$, the energies necessary to form the encounter complex from the reactants and to remove the products from it, and on the intrinsic barrier $\lambda_L/4$, the quantity related to the reorganization mentioned above. The preexponential factor Z is usually taken as a bimolecular collision frequency. The *Marcus* equation represents an analytical relationship between the rate constant and the reaction free energy and can thus be utilized as theoretical basis of free energy relations. If one assumes, as indicated in eq. (9) and (10), that the only isotope dependent quantity is $\lambda_L$ one arrives at the following expression for the KIE.

$$\frac{k_H}{k_D} = \exp\left\{\frac{\lambda_D - \lambda_H}{4k_B T}\left(1 - \frac{(\Delta G_0')^2}{\lambda_H\lambda_D}\right)\right\} \tag{12}$$

The shape of a free energy relation of the KIE according to eq. (12) is shown in Fig 3. It displays a maximum at $\Delta G_0' = 0$. This behaviour is now well documented by experimental data [1, 2]. It can be seen that the KIE in the exergonic or energonic region may be very small, in the vicinity of 1. Thus the usual reasoning that proton-transfer is only involved in a kinetic scheme if a KIE larger than 1 is found is in these cases not correct.

The MM in the version presented here can be only applied to the high temperature region. Within the theory of electron-transfer processes which bears a lot of analogies to proton-transfer there are recent attempts to design semiclassical versions of the MM by considering quantum effects only in the preexponential term, thus retaining to a certain extent the simplicity of the MM [8]. This would give a correct description of the temperature dependence for thermoneutral or nearly thermoneutral reactions. For exergonic or endergonic processes, however, it is necessary to consider quantum effects in the exponential term of eq. (9) too. The free energy dependence of the KIE can be described within the MM assuming isotope dependent values of the reorganization energies $\lambda_L$. According to our present knowledge it seems to be, however, very unlikely that the difference $\lambda_D - \lambda_H$ necessary to describe the experimental data correctly is in fact so large. Probably the isotope dependence of $\lambda_L$ is not more pronounced than for $\Delta G_0$, which was neglected in deriving eq. (12). If this is correct it has to be concluded that the KIE cannot be calculated in an appropiate way by the MM. This problem is discussed in more detail in section 3.2.

## 2.3. *Quantum-statistical mechanical model*

Within the QSM (frequently also called multiphonon theory) a chemical reaction is viewed as a particular relaxation process. This conception is especially interesting for photochemical reactions, because they are always accompanied by photophysical relaxation processes. Quantum-statistical mechanical rate theory should thus enable one principally to treat these competing processes by conceptually related models.

The idea to envisage a chemical reaction as a relaxation process was combined with a solvent model taking into account solvent fluctuations by *Dogonadze* and coworkers [9–11]. Within this model an intermolecular proton-transfer can be viewed in the following manner, Fig. 4. We adopt here the model of linear proton-transfer. In solution there is a

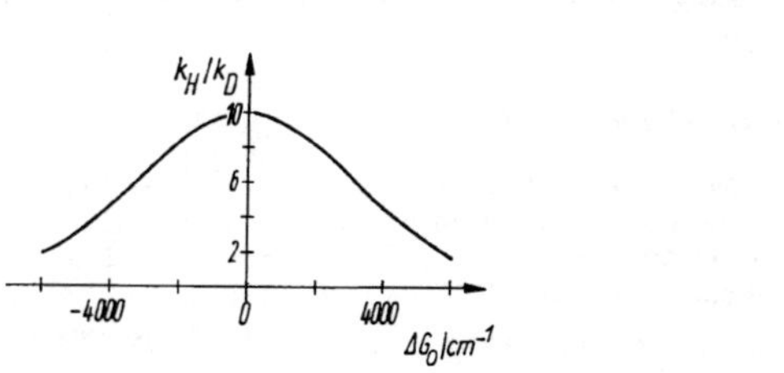


Fig. 3. Free energy dependence of the KIE according to the *Marcus* model $w_r = w_p$

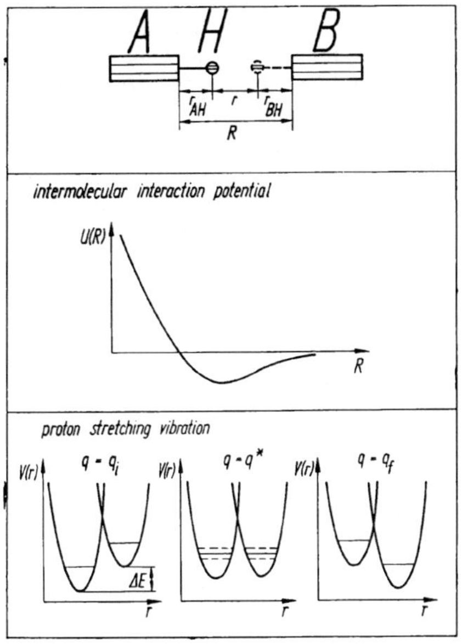


Fig. 4. Quantum-statistical mechanical model

distribution of pairs of reacting molecules with different intermolecular distances $R$. This distribution is governed by the intermolecular interaction potential $U(R)$. The whole system is subdivided into two subsystems, the quantum one (proton stretching vibration) and the classical one (all other intramolecular and the solvent modes). The energy levels of the proton stretching vibration are shifted by solvent fluctuations. This is indicated by different values of the effective solvent coordinate $q$. If the energy levels in the proton donor and acceptor have the same energy proton-transfer occurs by tunneling, indicated by the tunnel splitting effect. In order to arrive at the bimolecular rate constant $k_L$ one has to sum up over the vibrational states and to integrate over the intermolecular distance $R$. This results in the following rate expression [9]:

$$k_L = 4\pi \int_{R_{min}}^{\infty} dR\, R^2 \exp[-U_i(R)/k_B T]\, W_i^L(R) \tag{13}$$

$$W_i^L(R) = \sum_m p_{im}^L W_{im}^L(R) \tag{14}$$

$$W_{im}^L(R) = (\omega_{eff}/2\pi) \sum_n H_{mn}^L(R) \exp[-\sum_{k=0}^{n-1} 2\pi\gamma_{mk}^L(R)]$$
$$\times \exp\{-[E_s + \Delta E + U_f(R) - U_i(R) + E_{fn}^L - E_{im}^L]^2/4E_s k_B T \tag{15}$$

$$H_{mn}^L(R) = \frac{1 - \exp[-2\pi\gamma_{mn}^L(R)]}{1 - (1/2)\exp[-2\pi\gamma_{mn}^L(R)]} \tag{16}$$

$$\gamma_{mn}^L = \frac{V_c^2 S_{mn}^L(R)}{\hbar\omega_{eff}(k_B T E_s/\pi^3)^{1/2}\, 2\pi} \tag{17}$$

The symbols have the following meaning: $W_i^L$ – monomolecular rate constant; $W_{im}^L$ – state specific monomolecular rate constant; $U_i$, $U_f$ – intermolecular interaction potentials in the initial and final state of the reaction; $p_{mi}^L$ – population of the state $m$; $\omega_{eff}$ – medium frequency; $H_{mn}^L$ – transmission coefficient; $\gamma_{mn}^L$ – *Landau-Zener* parameter; $E_s$ – reorganization energy; $\Delta E$ – free energy of reaction; $E_{fn}^L$, $E_{im}^L$ – energies of the vibrational states; $S_{mn}^L$ – *Franck-Condon* factor; $V_c$ – electron resonance integral. The bimolecular rate constant $k_L$ may also be written as sum over state specific bimolecular rate constants $k_{mn}^L$, eq. (18).

$$k_L = \sum_m \sum_n k_{mn}^L \tag{18}$$

Provided all parameters are known $k_L$ can be calculated by numerical integration. Upon closer inspection of the integrand in eq. (13) it turns out, however, that it has a rather sharp maximum at some value of the intermolecular distance, the so-called transfer distance $R_{mn}^L$. Thus integration may be avoided by taking the value of the integrand at $R_{mn}^L$. Applying the usual approximation $R_{mn}^L = R_{00}^L$ eq. (13) transforms to eq. (19).

$$k_L = \exp[-U_i(R_{00}^L)/k_B T]\, W_i^L(R_{00}^L)\, 4\pi(R_{00}^L)^2 \Delta R \tag{19}$$

If furthermore only vibrational states with $m = n = 0$ are taken into account, which is justified in the vicinity of $\Delta E = 0$, one arrives at the approximate eq. (20).

$$k_L = 2\omega_{eff}(R_{00}^L)^2 \Delta R \exp[-U_i(R_{00}^L)/k_B T]\, H_{00}^L(R_{00}^L)$$
$$\times \exp\{-[E_s + \Delta E + U_f(R_{00}^L) - U_i(R_{00}^L)]^2/4E_s k_B T\} \tag{20}$$

Obviously this equation is completely analogous to the Marcus equation. Thus the MM can be viewed as an approximate version of the QSM. Due to this analogy the basic results are similar as far as quantum effects are negligible. This is true for the free energy dependence of $k_L$ (*Brønsted* relation) in the vicinity of $\Delta E = 0$. For the calculation of the free energy dependence of the KIE, however, the full expressions (eq. (13) or eq. (19)) have to be used and not eq. (20). Nevertheless the shape of the free energy plot of the KIE is qualitatively similar to the Marcus result, Fig. 3, though the physical origin is completely different. Within the QSM the free energy dependence of the KIE arises from the different thermal excitation of the proton and deuteron stretching vibration according to their different frequencies and is not the consequence of isotope dependent reorganization energies.

Finally it should be noted that the quantum-mechanical treatment is up to now confined to the nonadiabatic limit ($2\pi\gamma_{mn} \ll 1$). The resulting rate expression is identical with the semiclassical version. For the adiabatic case ($2\pi\gamma_{mn} \gg 1$) only a semiclassical treatment exists so far. Thus eq. (13) to (17) represent the quantum-mechanical result in a semiclassical language for the nonadiabatic case and the semiclassical result for the adiabatic case. The QSM is the most general of the models described. If one is interested in its application one runs, however, into difficulties. They are mainly related to the large number of parameter involved. Some of them can be taken from experiment ($T$, $\omega_{eff}$, $\Delta E$, vibrational energies, $E_s$ in an indirect way), others are unknown ($V_c$, $U(R)$). Nevertheless a detailed inowledge of the manner in which these quantities affect the rate constant combined with some known qualitative data may lead to meaningful results. Examples are presented in section 3. However, further studies on these parameters are highly desirable. This is especially true for the distance dependence of $V_c$ and the isotope and distance dependence of $E_s$. Concerning the basic structure of the theory the following developments would be important: consideration of more than one quantum degree of freedom and of weak interaction between solvent and solute. Though both of these problems were already discussed practical applications were not performed so far [10–12].

In 1984 a so-called golden-rule treatment of hydrogen-transfer was proposed [13]. The basic philosphy of this approach follows the QSM described above. The classical subsystem is, however, represented by only one low-frequency non-hydrogenic mode.

Finally it should be noted that the question usually posed whether tunneling is important or not has model dependent answers. Within the TM tunneling is important if $Q_t$ is larger than 1. For the MM in the version presented here this question makes no sense. Within the QSM tunneling is principally assumed to be operating for the quantum degree of freedom. Thus in this case the question has to be whether quantum effects play a role.

The present situation is characterized by intensive efforts to find out the ultimate consequences of the theoretical models available. In the majority of cases all the models seem to give qualitatively similar results. Thus a basic challenge to theory is to propose experiments for which the models give contradicting predictions.

## 3. Relation to experimental data

In this section theoretical results are compared to the corresponding experimental data. In all cases a short discussion on scope and limits of the different theoretical models will be given.

### 3.1. *Intermolecular interaction*

There is no doubt that bimolecular rate constants are affected by the interaction between the reactants. Within the QSM the corresponding potential $U(R)$ is explicitely involved in the rate expression. This model should thus be especially suitable for studying the influence of the shape of $U(R)$ on the rate constant. In the TM and MM rate expressions this effect can only be incorporated in a more or less indirect way by varying the barrier dimensions or the values of the work terms.

Quantitative data on the shape of the interaction potential between large molecules in solution are usually not available. On the other hand a few qualitative facts are well-known. It is, for example, certainly correct to say that the minimum of $U(R)$ is shifted to larger values of the intermolecular distance R for sterically hindered systems and to lower values of R for H-bonded molecules.

In the past it was usually assumed that $U_f = U_i$ and $\exp[-U_i/k_BT] = 1$. Applying the assumption $R^H_{mn} = R^D_{mn}$ the last term cancels out of the expression for the KIE. On the other hand there are indications that the influence of $U(R)$ is essential in some cases. Thus it was proposed in [14, 15] to utilize simple model potentials of the parabolic, *Lennard-Jones* or Morse type. This enables one to study in detail in which way variations in the potential parameters affect the KIE and if the assumptions mentioned above are justified. Combined with the qualitative knowledge on steric hindrance and H-bonding this might give a deeper insight into the role of $U(R)$. Of course the description of the interaction between two large molecules in a polar medium by an effective pair potential can be only of a very approximate nature.

The following results were obtained in [15]. The assumption $R^H_{mn} = R^D_{mn}$ is not justified for potential parameters of a realistic order of magnitude. This means that the term $\exp[-U_i/k_BT]$ has to be taken into account for the calculation of the KIE. The KIE increases with an increasing value of the intermolecular equilibrium distance (steric hindrance) and decreases with a decreasing value of the equilibrium distance (H-bonding). The result referring to steric hindrance is supported by a variety of experimental data [15, 16]. For example proton-transfer between 2-nitropropane and bases gives a KIE fo 10·3 for pyridine ($pK_a = 5{\cdot}2$) and 19·5 for 2,6-dimethylpyridine ($pK_a = 6{\cdot}7$) [17]. It is important to note that such a comparison requires constant values of all the other quantities affecting the magnitude of the KIE. This fact is occasionally overlooked, which results in wrong conclusions [18]. One of the main contributors to variations in the KIE is the free energy of reaction. Comparisons mentioned above can thus be only made for systems of at least nearly equal ΔpK values or even better if whole free energy relations are known. For proton-transfer between electronegative atoms with strong H-bonds such relations were not known until recently. Only data for individual reactions were available, among others also for excited electronic states [19–22]. The reported isotope effects were usually rather low, between 1 and 2. Within the last few years, however, free energy relations of this type were reported [23, 24]. They exhibit a maximum isotope effect between 3 and 5, considerably lower than for proton-transfer between C and O or N (7–11), and a more pronounced curvature of the plot. Assuming that the shape of these curves is exclusively determined by proton-transfer and not by a transition between rate-limiting proton-transfer and diffusion control this fact directs to low values of $E_s$ or $\lambda_L$. All these data are in agreement with the results of the QSM if one realizes that the expected small transfer distances in these systems give small reorganization energies.

The work terms of the MM can be identified with $U_i$ and $U_f$ of the QSM. The maximum value of the KIE at $\Delta G_0' = 0$, however, cannot be influenced by these terms, see eq. (12), but only by the difference $\lambda_D - \lambda_H$. According to our view the interpretation with the QSM is superior to the one within the MM.

In the TM it can be assumed that at room temperature and for parameters which give a KIE of the correct order of magnitude only the first term of eq. (3) or (5) has to be considered. In this case the only quantity affecting $Q^L_t$ and thus $k_H/k_D$ is the barrier frequency $\nu_L$ ($\nu_H = 2^{1/2}\nu_D$). In passing to sterically hindered systems both the width of the barrier ($2b_L$) and its height ($V^L_0$) should increase. The relative importance of these effects is crucial for the final result. An increase of the KIE is only obtained if the curvature of the parabola becomes more pronounced (higher barrier frequency). This means that the increase in $V^L_0$ must exceed the one in $2b_L$. To the best of our knowledge there are no theoretical studies of potential curves for proton-transfer between carbon and nitrogen or oxygen dealing with this problem.

### 3.2. *Free energy relations*

A great deal of the experimental material refers to proton-transfer between C and O or N with values of the KIE in the range between 1 and 11, provided further effects such as steric hindrance are not operating. This prompted of course the question to which origin the different values can be traced back.

In 1961 *Westheimer* showed that the isotope effect should be largest for a symmetrical transition state [25]. Later-on this idea was related to the free energy of reaction, more easily amenable to experimental investigations, by the Hammond postulate. This postulate states that exergonic reactions have reactant-like and endergonic reactions product-like transition states. Thus maximum isotope effects are to be expected for thermoneutral reactions. This behaviour is now clearly documented by experimental data [2, 3, 16, 26].

The MM and the QSM give also maximum isotope effects for $\Delta G_0' = 0$ or $\Delta E + U_f - U_i = 0$ [27–31]. The variation of $k_H/k_D$ with $\Delta G_0$ or $\Delta E$ with all other parameters kept fixed represents the so-called free energy relation of the KIE. Al already noted the shape of these relations is similar for the MM and the QSM although the physical origin is completely different. Thus to find out the more subtle differences between the two approaches a quantitative treatment is necessary. Within the MM the maximum KIE is determined by $\lambda_D - \lambda_H$ and the curvature by $\lambda_D\lambda_H$, eq. (12). Large values of $\lambda_L$ give a broad and flat curve and small values of $\lambda_L$ a steep fall-off. Within the QSM $E_s$ has a similar effect. The shape of the QSM plot is, however, also affected by other parameters. It becomes steeper by decreasing values of $V_e$ and increasing values of the proton-stretching frequency. Taking into account anharmonicity of this vibration would give a more flat curve as compared to the harmonic case [27, 28]. Moreover solving eq. (13) by numerical integration results in a slightly steeper curve [34]. The quantum-oscillations found for small values of $E_s$ are not discussed here.

In Fig. 5 MM and QSM results are compared using the QSM approach of [15, 29, 32] taking $\lambda_H = E_s$ and choosing the other QSM parameters in such a way as to produce a highly curved free energy plot. Nevertheless it can be seen from fig. 5 that the MM plot is generally more curved than the QSM one. Within a constant region of $\Delta G_0$ or $\Delta E$ this difference is the more pronounced the smaller $E_s = \lambda_H$ is. This result has interesting consequences. The value of $\lambda_H$ may be determined experimentally by fitting the *Brønsted* plot (lg $k_H/\Delta p$K) to a parabola. From the coefficient of the quadratic term $\lambda_H$ may be calculated. The corres-

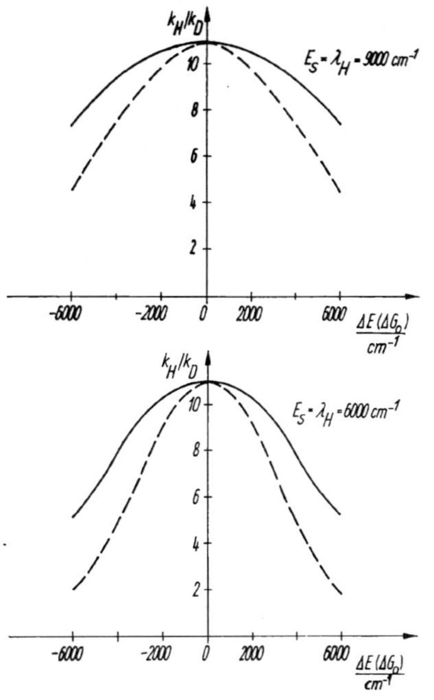


**Fig. 5.** Free energy dependence of the KIE calculated by the *Marcus* model (– – – –) and the quantum-statistical mechanical model (———)
$T$ = 300 K, $k_0 = r_{AH} + r_{BH}$ = 212 pm, $\nu_{eff} = 10^{12}\,s^{-1}$, $\tilde{\nu}_i^H = \tilde{\nu}_f^H$ = 3000 cm$^{-1}$, $\varepsilon$ = 30 cm$^{-1}$, $\sigma$ = 310 pm, $V_c$ = 10 cm$^{-1}$, numerical integration

ponding value is equal to $E_s$ as for not too exergonic or endergonic regions the MM and QSM rate expressions for $k_H$ are analogous, eq. (9–11, 20). Now $\lambda_H$ and $E_s$ can be used as input parameters for the free energy relation of the KIE. The theoretical curves obtained may in turn be compared to the corresponding experimental data. This procedure was for example applied to proton-transfer between 4-(4-nitrophenoxy)-butanone-2 and mainly oxygen bases [33]. The resulting theoretical *Marcus* plot for the KIE was too step and it was thus concluded that other effects should be operating (transition state solvation). The QSM gives, however, a more flat curve which is in better agreement with the experimental data. Thus other effects have not to be invoked. A systematic comparison of the free energy dependence of the KIE within the MM and the QSM is in progress [34].

For intramolecular excited state proton-transfer in some cases isotope effects were found (9-hydroxy-phenalenone [35], tropolone [36]) in others not (salicylideneaniline [37], o-hydroxyphenyl-s-triazine [38]). From the absence of the isotope dependence it was concluded by a few authors that the process observed did not involve the proton. Looking for differences between these examples it turns out that the former are thermoneutral processes, the latter, however, highly exergonic. Bearing in mind in which way the KIE depends on the reaction free energy the values observed are by no means surprising, fig. 3. Thus the conclusion that the proton is not involved in the latter cases is not correct.

The *Bell* approach of the TM gives also a maximum isotope effect for thermoneutral reactions [3]. For practical applications this approach was nearly not uptaken.

The maximum of the free energy relation of the KIE may be shifted into the endergonic or exergonic region if the work terms or $U_f$ and $U_i$ are different. Within the

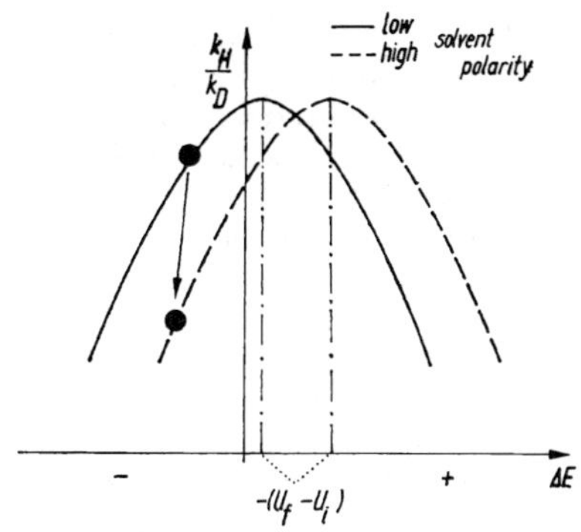


**Fig. 6.** Qualitative picture of the solvent effect on the free energy dependence of the KIE for reactions of the type $AH + B \rightleftharpoons A^- \cdot$ to $HB^+$

QSM such a shift may also be caused by different frequencies of the quantum degree of freedom in the initial and final state of the reaction. Usually the experimental data are widely scattered and display a rather broad maximum [2, 3, 26]. In these cases it is difficult to fix the location of the maximum. Recently a free energy plot of the KIE was reported for the reaction between carboxylic acids and a protonated cryptand [24].

$$LH^+ + HA \leftrightarrows LH^{2+} + A^- \tag{21}$$

It shows a rather sharp maximum at $\Delta pK = +1{,}2$. The rather low value of the maximum isotope effect ($k_H/k_D = 4$) is in agreement with the proposal of [15] for proton-transfer between electronegative atoms. The shift of the maximum into the endergonic region may be explained by the type of reaction (21). The interaction between the products is certainly stronger than between the reactants and this gives $U_f - U_i < 0$ at the transfer distance, Fig. 6.

### 3.3. *Temperature dependence*

The temperature dependence of the KIE has recently attracted much interest. This is probably due to the fact that the phenomenon of tunneling in chemical reactions is interrelated to it. It seems to be generally accepted now that non-linear *Arrhenius* plots are a consequence of tunneling. Usually the TM is used for describing this effect. Its basic results were presented in Fig. 2. The most direct evidence of tunneling is a non-linear *Arrhenius* plot. There are, however, only few examples for which a temperature range large enough to find such a behaviour was studied. In the overwhelming majority of cases the temperature region under study is so small that deviations from linearity cannot be detected. In these cases the values of $A_D/A_H$ and $E_a^D - E_a^H$ are used as criteria for tunneling. There is usually no difficulty to adjust the barier dimensions so as to fit the exeperimental data, see for example [39]. The question is, however, whether the data obtained are mere numbers or can be related to experimental properties. Nearly all examples for which a large temperature range was studied are either of the intramolecular type or solid state reactions. Most of these reactions were classified as H-atom and not proton-transfer though differentiating between these types seems to be not easy for intramolecular processes. Non-linear plots were found for the systems fluorenone/acridine [40], di- and hexahydrocarbazole [41], 2-methylacetophenone [42], tri-t-butylphenyl [43], dihydrophenanthrene/$O_2$ [44], dimethylglyoxime [45], methyl [46], tetraphenylporphyrine [47]. As already noted the Bell approach is frequently applied to intramolecular proton-transfer though in this case the assumption of a continuos distribution of states is certainly not justified. It is thus not surprising that in some cases difficulties were noticed in reproducing the experimental data [41]. The TM for a small number of states was applied to proton-transfer in tetraphenylporphyrine [47]. Very recently the so-called golden rule tetratment, a particular QSM approach, was applied to the temperature dependence [13]. The basic difference of this approach to the QSM described in 2·3. is that the classical subsystem is represented by one low-frequency non-protonic mode. The agreement between experiment and theory is in most cases satisfying though some discrepancies remain to remove.

Systematic studies of the temperature dependence by the QSM were not performed to date. Preliminary results, however, show that the values of $A_D/A_H$ and the difference of the apparent activation energies for H and D are strongly dependent on the frequency of the quatum mode. The experimental values for systems where tunneling is important according to the TM cannot be reproduced with $\tilde{\nu}_H = 3000\ cm^{-1}$ [27, 48].

The temperature dependence of the KIE has been proposed as a new mechanistic criterion [49]. It was suggested that non-linear transition states should have nearly temperature independent values of the KIE ($E_a^D - E_a^H \approx 0$) and $A_H/A_D > 2^{1/2}$. Though there are some indications that bent transition states should have smaller values of the KIE and smaller temperature coefficients than linear ones, anomalous high values of $A_H/A_D$ could not yet be rigirously derived from a theoretical model. On the contrary, it was shown by quantum-chemical calculations that there should be no marked difference in $A_H/A_D$ for linear and bent transition states [50].

### 3.4. *Solvent dependence*

For the proton-transfer reaction between 4-nitrophenylnitromethane and tetramethylguanidine a dramatic solvent dependence of the KIE was reported in a paper by *Caldin* and *Mateo* in 1975 [51]. Unusually large values of the KIE were found in non-polar solvents such as toluene or chlorobenzene ($k_H/k_D = 40-50$), while in more polar solvent such a acetonitrile or dichloromethane the KIE was much smaller ($k_H/k_D = 11-13$). The high values of the KIE were explained by a marked degree of tunneling. It was assumed that there should be a coupling between the motion of solvent molecules and of the proton/deuteron in polar solvents resulting in an increased effective mass of the transferring particle and thus a smaller isotope effect. Upon reinvestigating the same reactions *Rogne* pointed out that the deuteron transfer was complicated by a H-D exchange resulting in smaller values of the KIE than were previously reported. It was suggested from these results that there should be no marked effect of solvent polarity on the KIE [52]. *Kresge* and *Powell* studied the detritiation in the same system and found also, applying the *Swain-Schaad* relation, smaller values of the KIE [53]. On the other hand, *Caldin* et al. presented arguments, which tend to support their original results [54]. Very recently a marked solvent dependence of the KIE was found using the base 1,8-diazabicyclo[5.4.0]undec-7-ene (DBU), which has no exchangable proton. The KIE for the reaction with trinitrotoluene is 29.9 in 1,2-dichloroethane and 19.1 in acetonitrile [39]. The corresponding values for the reaction with 1-(4-nitrophenyl)-1-nitroethane are 12.0 in toluene and 7·8 in acetonitrile [55]. In both cases a decrease of the KIE by about 30% is found. These two examples provide ample evidence for a marked solvent dependence of the KIE even though the discussion on the tetramethylguanidine reaction has not yet finished.

Within the TM the experimental data were explained by assuming either solvent dependent effective masses or varying barrier dimensions. In [56] an interpretation of the solvent effect was given in terms of the QSM. This model should be especially useful for the treatment of solvent effects as solvent properties are explicitely involved in the rate expression. The solvent dependence of the proton-transfer rate constant and thus of the KIE originates from the way the parameters involved in the rate expression are affected by the solvent. It can be assumed that $\Delta E$, $E_s$ and $U(R)$ are the main contributors to the rate constant variation. The solvent dependence of $\Delta E$ is known experimentally for the examples mentioned above. $E_s$ can be assumed to increase with increasing solvent polarity. Neither variation in $\Delta E$ nor in $E_s$ alone can, however, account for the experimental facts. Of particular interest is thus the solvent dependence of $U(R)$. In all cases mentioned above proton-transfer transforms two neutral molecules into an ion-pair. By the same reasoning as invoked in 3.2. it can be concluded that $U_f - U_i < 0$ at the transfer distance. Moreover it is to be expected that the solvent dependence of $U_f(R)$ is more pronounced than for $U_i(R)$. This results in more negative values of $U_f - U_i$ at the transfer distance with increasing solvent polarity and thus in a shift of the maximum of the free energy plot into the endergonic region, fig. 6. In this way a marked decrease of the KIE in passing to more polar solvents can be explained by the QSM for the slightly exergonic regions under study. According to this result a marked solvent dependence of the KIE can be only expected if the intermolecular interaction potentials in the initial and final state of the reaction exhibit a different solvent dependence. More direct evidence of this view could be provided if the solvent dependence of complete free energy plots were known. Such experimental data, hoewver, are not available so far.

### 3.5. *Pressure dependence*

The pressure effect of the KIE was recently proposed as a further criterion for the identification of tunneling. For systems where tunneling was assumed to be operating a pressure effect could be observed, whereas in systems not associated with tunneling the KIE was almost unchanged with pressure [57]. In this respect the reaction between 2,4,6-trinitrotoluene and DBU has attracted particular interest. It was found that increasing the pressure from 1 to 1000 bar results in a decrease of the KIE from 19·1 to 16·9 and in an increase of the individual rate constants (solvent acetonitrile) [58, 59]. Moreover the quantities $E_a^D - E_a^H$ and $A_D/A_H$ decrease and the apparent activation energies $E_a^D$ and $E_a^H$ increase with increasing pressure. In [58, 59] an interpretation of these facts was presented in terms of the TM invoking either varying barrier dimensions or effective masses. In [58] the barrier width was increased and its height decreased with increasing pressure. This gives a smaller barrier frequency and thus a smaller isotope effect. In this way all the experimental data could be reproduced correctly except for the apparent activation energies. This was, however, not discussed in [58]. Moreover it is felt that the assumption of a broad and flat barrier for the high pressure case is not very likely. In [59] the effective masses were varied with pressure assuming unchanged barrier dimensions. Increasing effective masses result in decreased isotope effects. A more detailed comparison with other experimental data was not performed in [59]. Probably the increased effective masses will not result in increasing rate constants observed experimentally. Thus the theoretical analysis of the pressure dependence of the KIE requires further efforts. To date the QSM has not yet been applied to this phenomenon.

### 3.6. *Ultrafast proton-transfer*

Since the advent of laser techniques ultrafast processes can be investigated. For such processes the usual assumption that vibrational relaxation is faster than the competing reaction may be no longer justified. According to the QSM very fast proton-transfer may occur within systems having large negative reaction free energies, small reorganization energies and transfer distances and large electron resonance integrals. Moreover diffusion has to be excluded. These requirements can only be met by intramolecular proton-transfer or by intermolecular systems embedded in orgazined assemblies. It should, however, be noted that the term "ultrafast" is subject to varied interpretations embracing also fast intermolecular proton-transfer in solution for which vibrational relaxation is faster than the transfer of the proton. Provided proton-transfer is faster than vibrational relaxation the models described in section 2 have to be modified. In order to treat a non-Boltzmann distribution one has to use mode-specific models. Such models are the QSM and the TM for a small number of states, but not the MM and the *Bell* approach.

According to a stoachstic model proposed by *Jortner* for electron-transfer the rate expression has to be modified in the following manner [60]. The rate constant is given in a similar way as for the case of thermal equilibrium, but now the populations of the individual states are no longer time independent, eq. (22).

$$W_i^L(t) = \sum_m p_{im}^L(t)\, W_{im}^L \tag{22}$$

The populations are given according to eq. (23), where the $\varkappa_{mm'}^L$ are the vibrational relaxation rates.

$$dp_{im}^L(t)/dt = \sum_{m \neq m'} \varkappa_{m'm}^L p_{im'}^L(t) - \sum_{m \neq m'} \varkappa_{mm'}^L p_{im}^L(t) - W_{im}^L p_{im}^L(t) \tag{23}$$

The initial condition is

$$p_{im}^L(t = 0) = \delta_{mk} \tag{24}$$

where $k$ stands for the initially prepared state. There are two limiting cases.

$$W_{im}^L \ll \varkappa_{mm'}: \quad W_i^L = \sum_m p_{im}^L W_{im}^L \tag{25}$$

$$W_{im}^L \gg \varkappa_{mm'}: \quad W_i^L = W_{ik}^L \tag{26}$$

Tab. 1. Kinetic isotope effects and relative rate constants for a monomolecular proton-transfer reaction
($W_i^L$ – thermally averaged rate, $W_{im}^L$ – mode-spefific rate; transfer distance: 30 pm, $\tilde{\nu}_i^H - \tilde{\nu}_f^H = 3000\ cm^{-1}$, $\nu_{eff} = 10^{12}\ s^{-1}$, $E_s = 3000\ cm^{-1}$ $V_e = 10\ cm^{-1}$, $T = 300$ K)

$\Delta E = 0{\cdot}0\ cm^{-1}$
$W_i^H = 2{\cdot}56 \times 10^8\ s^{-1}$; $W_i^D = 4{\cdot}90 \times 10^7\ s^{-1}$; $W_i^H/W_i^D = 5{\cdot}32$

| $m$ | $W_{im}^H/s^{-1}$ | $W_{im}^D/s^{-1}$ | $W_{im}^H/W_{im}^D$ | $W_{im}^H/W_{i,m+1}^D$ |
|---|---|---|---|---|
| 0 | $2{\cdot}56 \times 10^8$ | $4{\cdot}86 \times 10^7$ | 5·26 | 0·03 |
| 1 | $3{\cdot}96 \times 10^{10}$ | $8{\cdot}42 \times 10^9$ | 4·70 | 0·60 |
| 2 | $7{\cdot}64 \times 10^{10}$ | $6{\cdot}58 \times 10^{10}$ | 1·16 | 1·61 |
| 3 | $1{\cdot}33 \times 10^{10}$ | $4{\cdot}72 \times 10^{10}$ | 0·28 | – |

$\Delta E = -4000\ cm^{-1}$
$W_i^H = 1{\cdot}38 \times 10^{10}\ s^{-1}$; $W_i^D = 7{\cdot}68 \times 10^9\ s^{-1}$; $W_i^H/W_i^D = 1{\cdot}79$

| $m$ | $W_{im}^H/s^{-1}$ | $W_{im}^D/s^{-1}$ | $W_{im}^H/W_{im}^D$ | $W_{im}^H/W_{i,m+1}^D$ |
|---|---|---|---|---|
| 0 | $1{\cdot}38 \times 10^{10}$ | $7{\cdot}68 \times 10^9$ | 1·79 | 0·20 |
| 1 | $1{\cdot}34 \times 10^{11}$ | $6{\cdot}67 \times 10^{10}$ | 2·00 | 2·73 |
| 2 | $9{\cdot}24 \times 10^9$ | $4{\cdot}90 \times 10^{10}$ | 0·19 | 0·36 |
| 3 | $3{\cdot}67 \times 10^{10}$ | $2{\cdot}56 \times 10^{10}$ | 1·43 | – |

Provided the proton-transfer rate constant is much smaller than the vibrational relaxation rate the thermally averaged rate constant results, eq. (25). If, however, $W_{im}^L$ is much larger than $\varkappa_{mm'}^L$, only proton-transfer from the initially prepared state $k$ occurs, eq. (26). More important is probably the intermediate range. In this case the kinetic equations have to be solved numerically, which is usually impossible due to lacking information on the corresponding processes. To date our knowledge on proton-transfer processes is both from the experimental and theoretical point of view nearly exclusively confined to the limiting case of eq. (25). Recently it was definitely claimed that for 2-(2-hydroxyphenyl)benzothiazole the rate of the excited-state intramolecular proton-transfer should exceed the rate of vibrational relaxation by at least one order of magnitude [61]. According to our view one cannot be sure that the assumed value of the vibrational relaxation rate of $10^{11}\ s^{-1}$ is indeed correct. Nevertheless theoretical investigations should be useful to learn which changes are to be expected as compared to the thermally averaged case.

In Tab. 1 mode specific rate constants obtained by the QSM are presented. The TM for a small number of states gives similar results. In passing to higher excited vibrational states the rate constants increase and slow then down again, depending on the values of the reaction free energy. The second result refers to the KIE. There is certainly no reason to assume that the initially prepared states have the same quantum numbers for H and D. In this case unusual large or even inverse isotope effects may occur. If the quantum numbers are equal or differ only by 1 the KIE given in the table are obtained and even in this case inverse isotope effects arise. The occurrence of such anomalous effects may thus be a strong indication of proton-transfer faster than vibrational relaxation. However, this effect will be probably not so pronounced in the intermediate case.

## 4. Conclusions

According to our view it is highly desirable to make further attempts to find out the ultimate consequences of the different theoretical models. Unfortunately these models give qualitatively similar results for most of the experimental data available at present. Both theoretical predictions and new experimental facts could thus stimulate the development of the theory of proton-transfer reactions and the kinetic isotope effect.

Received June 28, 1985

## References

[1] *E. Buncel, C. C. Lee* (eds.), Isotopes in Organic Chemistry, Vol. 2, Isotopes in Hydrogen Transfer Processes. Amsterdam: Elsevier Scientific Publishing Company 1976

[2] *R. P. Bell*, The Proton in Chemistry, 2nd ed.; London- Chapman and Hall 1973

[3] *R. P. Bell*, The Tunnel Effect in Chemistry. London: Chapman and Hall 1980
[4] *R. T. Skodje, D. G. Truhlar*, J. Phys. Chem. **85** (1981) 624
[5] *H. H. Limbach, J. Hennig, D. Gerritzen, H. Rumpel*, Faraday Disc. Chem. Soc. **74** (1982) 229
[6] *S. J. Formosinho*, J. Chem. Soc., Faraday Trans. 2, **70** (1974) 605
[7] *R. A. Marcus*, J. Phys. Chem. **72** (1968) 891
[8] *B. S. Brunschwig, J. Logan, M. D. Newton, N. Sutin*, J. Am. Chem. Soc. **102** (1980) 5798
[9] *E. D. German, A. M. Kuznetsov, R. R. Dogonadze*, J. Chem. Soc., Faraday Trans. 2, **76** (1980) 1128
[10] *R. R. Dogonadze, A. M. Kuznetsov*, Itogi Nauki Tekh., Fiz. Khim., Kinetika, Tom 2, Moskva: Viniti 1973
[11] *J. Ulstrup*, Charge Transfer Processes in Condensed Media. Berlin: Springer 1979
[12] *A. M. Kuznetsov*, J. Electroanal. Chem. **151** (1983) 227
[13] *W. Siebrand, T. A. Wildman, M. Z. Zgierski*, J. Am. Chem. Soc. **106** (1984) 4083
[14] *E. D. German, A. M. Kuznetsov*, J. Chem. Soc., Faraday Trans. 2, **77** (1981) 2203
[15] *J. Sühnel, K. Gustav*, Chem. Phys. **70** (1982) 109
[16] *L. Melander, W. H. Saunders*, Jr., Reaction Rates of Isotopic Molecules. New York: Wiley 1980
[17] *R. P. Bell, D. M. Goodall*, Proc. Roy. Soc. **A 294** (1966) 273
[18] *V. Gold, R. A. Lee*, J. Chem. Soc., Chem. Commun. **1984**, 1032
[19] *H. Shizuka, K. Tsutsumi, H. Takeuchi, I. Tanaka*, Chem. Phys. Lett. **62** (1979) 408
[20] *K. Tsutsumi, H. Shizuka*, Z. Phys. Chem. NF **122** (1980) 129
[21] *K. Tsutsumi, S. Sekiguchi, H. Shizuka*, J. Chem. Soc., Faraday Trans. 1, **78** (1982) 1087
[22] *K. Tsutsumi, H. Shizuka*, Z. Phys. Chem. NF **111** (1978) 120
[23] *A. J. Kresge*, Pure Appl. Chem. **53** (1981) 189
[24] *B. G. Cox, N. van Truong, H. Schneider*, J. Chem. Soc., Perkin Trans. **2** (1983) 515
[25] *F. H. Westheimer*, Chem. Rev. **61** (1961) 265
[26] *E. F. Caldin, V. Gold* (eds.), Proton Transfer Reactions. London: Chapman and Hall 1975
[27] *N. Brüniche-Olsen, J. Ulstrup*, J. Chem. Soc., Faraday Trans. 1, **75** (1979) 205
[28] *E. D. German, A. M. Kuznetsov*, J. Chem. Soc., Faraday Trans. 1, **77** (1981) 397
[29] *J. Sühnel, K. Gustav*, Chem. Phys. **87** (1984) 179
[30] *M. M. Kreevoy, S. W. Oh*, J. Am. Chem. Soc. **95** (1973) 4805
[31] *A. J. Kresge*, J. Am. Chem. Soc. **102** (1980) 7797
[32] *J. Sühnel*, Wiss. Z. Friedrich-Schiller-Univ. Jena, Math.-Naturw. R., **34** (1985) 685
[33] *D. J. Hupe, E. R. Pohl*, J. Am. Chem. Soc. **106** (1984) 5634
[34] *J. Sühnel*, in preparation
[35] *R. Rossetti, R. C. Haddon, L. E. Brus*, J. Am. Chem. Soc. **102** (1980) 6913
[36] *R. Rossetti, L. E. Brus*, J. Chem. Phys. **73** (1980) 154
[37] *P. F. Barbara, P. M. Rentzepis, L. E. Brus*, J. Am. Chem. Soc. **102** (1980) 2786
[38] *H. Shizuka, K. Matsui, Y. Hirata, I. Tanaka*, J. Phys. Chem. **81** (1977) 2243
[39] *N. Sugimoto, M. Sasaki, J. Osugi*, J. Phys. Chem. **86** (1982) 3418
[40] *J. B. Colpa, B. Prass, D. Stehlik*, Chem. Phys. Lett. **107** (1984) 469
[41] *K. H. Grellmann, U. Schmitt, H. Weller*, Chem. Phys. Lett. **88** (1982) 40
[42] *K. H. Grellmann, H. Weller, E. Tauer*, Dhe. Phys. Lett. **95** (1983) 195
[43] *G. Brunton, D. Griller, L. R. C. Barclay, K. U. Ingold*, J. Am. Chem. Soc. **98** (1976) 6803
[44] *A. Bromberg, K. A. Muszkat, E. Fischer, F. S. Klein*, J. Chem. Soc., Perkin. Trans. **2** (1972) 588
[45] *K. Toriyama, K. Nunone, M. Iwasaki*, J. Am. Chem. Soc. **99** (1977) 5823
[46] *R. J. LeRoy, H. Murai, F. Williams*, J. Am. Chem. Soc. **102** (1980) 2325
[47] *J. Hennig, H.-H. Limbach*, J. Chem. Soc., Faraday Trans. 2, **75** (1979) 752
[48] *J. Sühnel*, unpublished results
[49] *H. Kwart*, Acc. Chem. Res. **15** (1982) 401
[50] *B. Anhede, N.-A. Bergman*, J. Am. Chem. Soc. **106** (1984) 7634
[51] *E. F. Caldin, S. Mateo*, J. Chem. Soc., Faraday Trans. 1, **71** (1978) 1876
[52] *O. Rogne*, Act. Chem. Scand. **A 32** (1978) 559
[53] *A. J. Kresge, M. F. Powell*, J. Am. Chem. Soc. **103** (1981) 201
[54] *E. F. Caldin, S. Mateo, P. Warrick*, J. Am. Chem. Soc. **103** 1981 202
[55] *K. T. Leffek, P. Pruszynski*, Can. J. Chem. **60** (1982) 1692
[56] *J. Sühnel, K. Gustav*, Chem. Phys. 88 (1984) 391
[57] *N. S. Isaacs. K. Javaid, E. Rannala*, J. Chem. Soc., Perkin Trans. **2** (1978) 709
[58] *N. Sugimoto, M. Sasaki, J. Osugi*, J. Am. Chem. Soc. **105** (1983) 7676
[59] *N. Sugimoto, M. Sasaki, J. Osugi*, J. Chem. Soc., Perkin Trans. **2** (1984) 655
[60] *J. Jortner*, J. Am. Chem. Soc. **102** (1980) **6676**
[61] *K. Ding, S. J. Courtney, A. J. Strandjord, S. Flem, D. Friedrich P. F. Barbara*, J. Phys. Chem. **87** (1983) 1184

Isotopenpraxis 22 (1986) 3, S. 80

## Buchbesprechung

### Die Schilddrüse

### Aktuelle funktionsmorphologische Diagnostik. Therapie-Schemen

*H. Deckart. Hrsg.: Isocommerz GmbH, Berlin 1984. 175 S., 57 Abb., 32 Tab., 188 Lit. (deutsch/engl.; Schriftenreihe Anwendung von Isotopen und Kernstrahlungen in Wissenschaft und Technik, Heft 14)*

In Fortführung der 1972 erschienenen (Heft 4 der Schriftenreihe), auf nuklearmedizinische Untersuchungsverfahren ausgerichteten Monographie hat der Autor entsprechend der im vergangenen Jahrzehnt stattgefundenen Entwicklung eine weitergefaßte funktionsmorphologische Schilddrüsendiagnostik in konzentrierter und informativer Form zusammengefaßt. Dadurch werden nicht nur Stand und Tendenzen der Radiotraceranwendung aktualisiert ($^{99m}TcO_4^-$-Aufnahmetest, radiometrische Hormon- und Proteinbestimmungen, Bestimmung schilddrüsenspezifischer Antikörper, Therapie mit $^{131}I^-$), sondern die zum modernen Spektrum an Untersuchungsmethoden gehörenden nicht auf Radionuklidanwendung beruhenden Verfahren einbezogen und eingeordnet. In 17 Abschnitten werden folgende Teilgebiete abgehandelt: Grundlagen, morphologische Diagnostik, *in-vivo*-Funktionsdiagnostik, *in-vitro*-Funktionsdiagnostik, Untersuchungsverfahren zur Bestimmung schilddrüsenspezifischer Antikörper, Klinik und Diagnostik der Thyreopathien, Medikamenteneinfluß auf den Schilddrüsenhormonstoffwechsel, extrathyreoidale Erkrankungen und Schilddrüsenhormonstoffwechsel, rationelles Diagnostik-Stufenprogramm, Therapie von Thyreopathien, Radioiod-Therapie.

Die Monographie ist in erster Linie für den Mediziner von besonderem Interesse, dürfte aber auch für den medizinisch orientierten Naturwissenschaftler eine wertvolle moderne und durch aktuelle Literaturangaben bereicherte Darstellung der speziellen Thematik sein.

***B. Johannsen***

Isotopenpraxis 22 (1986) 3, S. 81 bis 87

# Zur anorganischen Chemie des monovalenten Astats

*R. Dreyer, I. Dreyer, W. Doberenz, S. Fischer*

(Technische Universität Dresden, Sektion Chemie, Wissenschaftsbereich Struktur- und Radiochemie)[1])

*W. A. Chalkin, F. Rösch*

(VIK Dubna, Laboratorium für Kernprobleme, Sektor Radiochemie)[2])

*Es wird ein Überblick zum gegenwärtigen Stand der Chemie des Astats im Oxidationszustand +1 gegeben. Entsprechend der Stellung des Astats im Periodensystem der Elemente besitzt At(I) „metallischen" Charakter. Dieser widerspiegelt sich in der Existenz eines einfach geladenen Kations $At(\theta)^+$ in saurer wäßriger Lösung sowie im Komplexbildungsverhalten dieses Ions. $At(\theta)^+$ bildet mit Halogeniden ($Cl^-$, $Br^-$, $I^-$) und Pseudohalogeniden ($SCN^-$, $CN^-$, $C(CN)_3^-$, $N_3^-$) anionische und mit Thioharnstoff, dessen Derivaten und N-Acylthioharnstoffen kationische Verbindungen, die im wäßrigen Milieu zum Großteil sehr beständig sind. Die Ermittlung der günstigsten Synthesebedingungen, der Zusammensetzung, der Existenz- und Stabilitätsbereiche der At(I)-Komplexe sowie die Abschätzung der Bildungskonstanten erfolgte hauptsächlich mittels Elektromigration in freier Elektrolytlösung.*

*There is given a review on the level of astatine chemistry in oxydation state +1. According to the position of astatine in the periodic system At(I) has „metallic" properties. They may be seen in the existence of similar changed cation $At(\theta)^+$ in acid water solutions and in complex formation reactions with halogenides ($Cl^-$, $Br^-$, $I^-$), pseudohalogenides ($SCN^-$, $CN^-$, $C(CN)_3^-$, $N_3^-$) to anionic forms and with thioureas its derivates and N-acylthioureas to cationic forms which are almost very stable in water solutions. The information on optimal synthese conditions, composition, existence and stability area of At(I)-complexes and on the calculation of stability constants mainly based on electromigration in free electrolyt solutions.*

## Keywords

astatine cation; astatine complexes; astatine 211; chemical properties; reviews;

## Einleitung

In den letzten fünf Jahren sind die Kenntnisse über den physiko-chemischen Zustand und das chemische Reaktionsverhalten des Astats im Oxidationszustand +1 rasch angewachsen. Dafür gibt es vor allem zwei Gründe. Bei Verwendung von $^{211}At$ sind auch die anorganischen Verbindungen des Elements als potentielle Radiotherapeutika von Interesse. In einer Reihe von Kernforschungszentren werden dehsalb verstärkt Arbeiten mit Astat in pharmazeutischer und biologisch-medizinischer Richtung durchgeführt. Des weiteren sind Untersuchungen der metallischen Eigenschaften des Astats von Bedeutung für die Voraussage des physikalisch-chemischen Verhaltens der „stabilen superschweren" Elemente. Eka-Astat liegt mit der Ordnungszahl 117 im Bereich der hypothetischen Stabilitätsinsel um das Eka-Blei und sollte Eigenschaften eines typischen Metalls besitzen.

In der vorliegenden Arbeit wird eine Übersicht über den gegenwärtigen Stand der anorganischen Chemie des Astats (I) gegeben. Der Schwerpunkt liegt dabei auf den seit 1979 erzielten Ergebnissen.

## 1. Das einfach positive Astation $[At(H_2O)_n^+]$

Kationen des Astats erhält man in sauren wäßrigen Lösungen durch Oxidation von $At^-$ oder $At^0 \cdot H_2O$ mit $Cr_2O_7^{2-}$, $S_2O_8^{2-}$ oder $XeF_4$ [1, 2]. Als $At^0 \cdot H_2O$ bezeichnen wir dabei die Astatform, die durch Destillation des Astats von einem Silberadsorber in Wasser erhalten wird. Ihr chemischer Zustand ist derzeit nicht eindeutig bekannt Der überwiegende Teil der bisher vorliegenden Untersuchungen konzentriert sich auf das Verhalten des Astats in sauren bichromathaltigen wäßrigen Lösungen. Eindeutige Aussagen über das durch Oxidation mit $S_2O_8^{2-}$ und $XeF_4$ in wäßriger Lösung gebildete Kation des Astats stehen noch aus, so daß auf dessen Behandlung nicht weiter eingegangen wird.

Im Gegensatz zu den anderen Halogenen ist das durch Oxidation mit $Cr_2O_7^{2-}$ erhaltene Astat-Kation in sauren Lösungen stabil. Seine Ladungszahl $Z$ wurde über Ionenaustauschuntersuchungen abgeschätzt. An monofunktionellen Sulfokationiten vom Typ Dowex 50 × 8 sowie an Aminex A7 erhielt man unabhängig vom verwendeten Säureanion bis zur Säurekonzentration von 1 mol · $l^{-1}$ einen linearen Zusammenhang zwischen dem Logarithmus des Verteilungskoeffizienten $D$ und der $H^+$-Konzentration [3, 4]. Die Ladung ergab sich hieraus zu +1. Weiterführende Untersuchungen seiner Eigenschaften zeigten, daß sich das Astat-Kation häufig anders als einfach positiv geladene Metallionen verhält. Es lag die Vermutung nahe, daß es sich um ein zusammengesetztes Ion der Form $[At(H_2O_n]^+$ handelt, und es erhielt die Bezeichnung $At(\theta)^+$ [5]. Aus heutiger Sicht sind für dieses Kation drei Eigenschaften hervorzuheben:

1. Sein Adsorptionsverhalten unterscheidet sich signifikant von dem anderer einwertiger Metallionen.

Anschriften: [1]) DDR-8027 Dresden, Mommsenstraße 13;
[2]) UdSSR 141980 Dubna, Vereinigtes Institut für Kernforschung

2. Seine Ionenbeweglichkeit ist vom pH-Wert abhängig und unter vergleichbaren Bedingungen geringer als die einwertiger Kationen wie $Cs^+$ oder $Tl^+$.
3. Es besitzt eine ausgeprägte Tendenz zur Komplexbildung, die sich mit der des $Ag^+$ und $Cu^+$ vergleichen läßt.

Eine ausführliche Zusammenfassung zum Adsorptionsverhalten des $At(0)^+$ ist in [5,18] gegeben. Ein charakteristisches Beispiel ist die Sorption von $At(0)^+$ aus $NaClO_4$-$HClO_4$-Lösungen konstanter Ionenstärke an Sulfokationiten. Man beobachtete ein Anwachsen der Verteilungskoeffizienten des $At(0)^+$ an Dowex 50 × 80 auf 100 bei abnehmender Säurestärke ($C_{H^+} = 0{,}2 - 1{,}0$ mol · l$^{-1}$). Für $Tl^+$ hingegen waren die Verteilungskoeffizienten unabhängig von der Acidität der Lösungen. Weiterhin nahmen die Verteilungskoeffizienten des $At(0)^+$ mit steigender Temperatur (Intervall 283–323 K) um mehr als die Hälfte ab. Die daraus berechnete Sorptionsenthalpie beträgt für das $At(0)^+$ 13,8 KJ/mol, die für das $Tl^+$ hingegen nur 6,3 KJ/mol [3]. Wie in [5,18] zusammengefaßt, läßt sich dieses Verhalten des $At(\theta)^+$ am besten mit der Bildung eines Aqueokomplexes der Zusammensetzung $[At(H_2O)_n]^+$ erklären. Derartige Komplexe werden für Silber [6] und Iod [7] beschrieben sowie verschiedentlich für das Hydroxonium-Ion [8] angenommen.

Einen wesentlichen Beitrag zur Charakterisierung des $At(\theta)^+$ erbrachte die Bestimmung seiner Ionenbeweglichkeit in freien Elektrolyten in Abhängigkeit vom pH-Wert (s. Tab. 1) [9]. Deutlich ausgeprägt ist die Abnahme der mittleren Ionenbeweglichkeit $u_m$ des Astat-Kations mit der Zunahme des pH-Wertes und der Einfluß der Ionenstärke [10]. Ausgehend von der Gleichung

$$[At(H_2O)_n]^+ \rightleftharpoons [At(H_2O)_{n-1}OH] + H^+; \quad K_s$$

läßt sich über die Elektromigrationsbeziehung

$$u_m = \frac{u_{At(\theta)^+}}{1 + \frac{K_s}{a_{H^+}}}$$

die Säurekonstante (Deprotonierungskonstante) für den Aquokomplex des At(I) berechnen (Tab. 1).

Der $K_s$-Wert liegt in der gleichen Größenordnung wie der für die protonierte hypoiodige Säure $[I(H_2O)]^+$ abgeschätzte Wert von $K_s \simeq 0{,}03$ [11] und ist höher als aus dem Gang der Halogene für Astat zu erwarten war.

Die Ergebnisse lassen den Schluß zu, daß in $Cr_2O_7^{2-}$-haltigen Elektrolyten Astat ausschließlich als Aquokomplex in der Oxidationsstufe +1 vorliegt. Die nachfolgend angeführten Ergebnisse zur Komplexbildung des $At(\theta)^+$ machen eine Zusammensetzung von $[At(H_2O)_2]^+$, wahrscheinlich.

Die hier gegebene Charakterisierung des physikochemischen Zustandes des Astat(I)-Kations findet ihre Fortsetzung bei Betrachtungen zur organischen Chemie des Astats. Ergebnisse zur elektrophilen Astatierung von aromatischen Verbindungen stützen die hierbei gewonnenen Resultate zum chemischen Zustand des $At(0)^+$ [12].

## 2. Das Assoziations- und Komplexbildungsverhalten von At(I)

### 2.1. *Halogenidkomplexe*

Die Interhalogenverbindungen des Astats sind am umfassendsten untersucht. Nachweise zur Bildung von AtCl, AtBr, AtI erfolgten in der Gasphase. Die AtX-Verbindungen wurden durch direkte Umsetzung von Astat mit Makromengen an Chlor, Brom und Iod erhalten. Ihre Identifizierung erfolgte massenspektroskopisch [13] bzw. durch Gasthermochromatographie [14–16].

Erste Untersuchungen zu Interhalogenverbindungen des Astats in wäßrigen Lösungen führte *E. H. Appelmann* durch [17]. Vom gleichen Autor wurden auf der Grundlage von Extraktionsuntersuchungen die Gleichgewichtskonstanten für die jeweiligen Reaktionen ermittelt (s. Tab. 2). Weitere Resultate zur Kenntnis der Interhalogenverbindungen bis zum Jahr 1977 sind in [5,18] zusammengefaßt. In diesem Beitrag sind auch die ersten Ergebnisse zum Komplexbildungsverhalten des $At(\theta)^+$ in wäßrigen Lösungen, die mittels Adsorptionsmessungen erhalten wurden, angegeben (s. Tab. 2).

Die Anwendung der Elektromigrationsmethode machte es möglich, das Komplexbildungsverhalten des $At(\theta)^+$ in halogenidhaltigen Elektrolyten unterschiedlicher Zusammensetzung ausführlich zu untersuchten [19–21]. Auf diesem Wege konnten Synthesebedingungen, Komplexbildungskonstanten und die Stabilitätsbereiche für die Astathalogenidkomplexe $AtCl_2^-$, $AtBr_2^-$ und $AtI_2^-$ ermittelt werden.

Es wurde nachgewiesen, daß in sauren halogenidhaltigen Lösungen die Synthese der Halogenidkomplexe des Astats sowohl ausgehend vom $At(0)^+$, von $At^0 \cdot H_2O$ und auch über $At^-$ möglich ist [22]. In Abb. 1 sind Ergebnisse zur Elektromigration von Halogenidkomplexen des Astats in Abhängigkeit von der Ligandenkonzentration zusammengefaßt (Abb. 1).

Tab. 1. Mittlere Ionenbeweglichkeiten $u_m$ des $At(\theta)^+$ für verschiedene Säurekonzentrationen bei 298 K und daraus berechnete Kationenbeweglichkeiten $u_{At(\theta)^+}$ sowie Säurekonstanten $K_s = a_{AtOH} \cdot a_{H^+}/a_{At(\theta)^+}$

| Elektrolyt | | | | | | | | | | | $u_{(At\theta)^+}$ $10^{-4}$ cm²/V · s | $K_s$ mol/l | Lit. |
|---|---|---|---|---|---|---|---|---|---|---|---|---|---|
| $(0{,}4 - y)$ M $HClO_4$ − $y$ M $NaClO_4$ − $10^{-4}$ M $K_2Cr_2O_7$ ($0 \leq y \leq 0{,}4$; $I = 0{,}4$ mol/l) | pH | 1,68 | 1,56 | 1,42 | 1,21 | 1,09 | 0,93 | 0,81 | 0,71 | 0,63 | 3,08 | 0,032 ±0,005 | [9] |
| | $u_m$ | 1,17 | 1,34 | 1,67 | 2,38 | 2,41 | 2,43 | 2,45 | 2,64 | 2,67 | | | |
| $(0{,}04 - y)$ M $HClO_4$ − $y$ M $NaClO_4$ − $5 \cdot 10^{-3}$ M $Na_2Cr_2O_7$ ($0 \leq y \leq 0{,}03$); $I \simeq 0{,}05$ mol/l) | pH | 2,00 | 1,90 | 1,75 | 1,70 | 1,65 | 1,60 | 1,55 | 1,40 | – | 4,23 | 0,035 ±0,015 | [10] |
| | $u_m$ | 0,78 | 0,83 | 1,18 | 1,50 | 1,87 | – | 2,13 | 2,69 | – | | | |

Tab. 2. Individuelle Stabilitätskonstanten $K_1$ und $K_2$ für die Bildung von AtX und $AtX_2^-$ aus $At(\theta)^+$
IA – Ionenaustausch; EM – Elektromigration in freier Elektrolytlösung; EX – Extraktion
$u_i$ in $10^{-4}$ cm²/V · s

| $X^-$ | wäßrige Lösung | $\vartheta$ in °C | $K_1$ in l/mol | $K_2$ in l/mol | Methode | Bemerkung | Lit. |
|---|---|---|---|---|---|---|---|
| $Cl^-$ | 0,3 M $HClO_4$ – $y$ M HCl ($2 \cdot 10^{-3} \leqq y \leqq 0{,}01$) | | 1300 | 200 | | | |
| | 0,5 M $HClO_4$ – $y$ M HCl ($2 \cdot 10^{-3} \leqq y \leqq 6 \cdot 10^{-3}$) | 20 | 1600 | 156 | IA | | [5] |
| | 0,5 M $HNO_3$ – $y$ M HCl ($2 \cdot 10^{-3} \leqq y \leqq 6 \cdot 10^{-3}$) | | 700 | 355 | | | |
| | (0,15 – $y$) M $HClO_4$ – $y$ M HCl – $10^{-3}$ M $K_2Cr_2O_7$ ($10^{-6} \leqq y \leqq 0{,}15$) ($I \simeq 0{,}15$ mol/l) | 25 | 500<br>500 | (500) Kurve (2) in Abb. 1<br>200 Kurve (1) in Abb. 1 | EM | s. Abb. 1<br>$u_{At(\theta)^+} = +3{,}7$<br>$K_s = 0{,}032$<br>$u_{AtCl_2^-} = -3{,}75$ Kurve (2)<br>$u_{AtCl_2^-} = -3{,}90$ Kurve (1) | [21] |
| $Br^-$ | 0,063 M $HClO_4$ – (0,40 – $y$) M $NaClO_4$ – $y$ M NaBr – $I_2$ ($1 \cdot 10^{-4} \leqq y \leqq 0{,}9$) | 21–25 | – | 320 | EX | in $CCl_4$; $I_2$ gelöst ($1 \cdot 10^{-6}$ M $\leqq c_{I(CCl_4)}$ $\leqq 0{,}08$) | [17] |
| | 0,02 M $HClO_4$ – (0,03 – $y$) M $NaClO_4$ – $y$ M NaBr ($5 \cdot 10^{-4} \leqq y \leqq 0{,}03$; I $\simeq 0{,}05$) | 25 | – | 250 | EM | s. Abb. 1<br>$u_{AtBr_2^-} = -4{,}7$ | [20] |
| | | ~ 28 | | 310 | | $u_{AtBr_2^-} = -4{,}8$ | [10] |
| $I^-$ | ~ $10^{-3}$ M $HClO_4$ – $y$ M NaI ($2 \cdot 10^{-5} \leqq y \leqq 0{,}1$) | 21–25 | – | 2000 | EX | in $CCl_4$; $I_2$ gelöst ($1 \cdot 10^{-6}$ M $\leqq c_{I_2(CCl_4)}$ $\leqq 0{,}08$) | [17] |
| | 0,02 M $HClO_4$ – (0,03 – $y$) M $NaClO_4$ – $y$ M KI | 25 | – | 1600 | EM | s. Abb. 1<br>$u_{AtI_2^-} = -4{,}15$ | [20] |
| | ($10^{-4} \leqq y \leqq 0{,}03$; I $\simeq 0{,}05$) | ~ 28 | | 1680 | | $u_{AtI_2^-} = -4{,}30$ | [10] |
| | (0,05 – $y$) M $NaClO_4$ – $y$ M KI | 18 | – | 2750 | | $u_{AtI_2^-} = -3{,}63$ | [10] |
| | ($10^{-4} \leq y \leq 0.05$; $I \simeq 0.05$) | 25 | – | 2040 | EM | $u_{AtI_2^-} = -4{,}12$ | |
| | | 33 | – | 1530 | | $u_{AtI_2^-} = -4{,}90$ | |
| | | 40 | – | 1290 | | $u_{AtI_2^-} = -5{,}48$ | |
| $SCN^-$ | 0,02 M $HClO_4$ – (0,03 – $y$) M $NaClO_4$ – $y$ M $NH_4SCN$ ($10^{-4} \leqq y \leqq 0{,}03$; $I = 0{,}05$) | 25 | – | 400 | EM | s. Abb. 1<br>$u_{At(SCN)_2^-} = -3{,}7$ | [26] |
| $NO_3^-$ | (1 – $y$) M $HClO_4$ – $y$ M $HNO_3$ ($0{,}1 \leqq y \leqq 1$) | 20 | 1,35 | – | IA | | [5] |
| $SO_4^{2-}$ | (1 – $y$) M $HClO_4$ – $y$ M $H_2SO_4$ ($0{,}1 \leqq y \leqq 1$) | 20 | 1,75 | 1,77 | IA | | [5] |
| $Cr_2O_7^{-2}$ | 1 M $HClO_4$ – $y$ M $H_2Cr_2O_7$ ($10^{-3} \leqq y \leqq 0{,}05$) | 20 | 22 | – | IA | | [5] |
| | 0,04 M $HClO_4$ – (0,03 – $y$) M $NaClO_4$ – $y$ M $Na_2Cr_2O_7$ $5 \cdot 10^{-3} \leqq y \leqq 0{,}02$; $I = 0{,}1$) | 25 | 22 | – | EM | | [10] |
| $S_2O_8^{2-}$ | 0,04 M $HClO_4$ – (0,03 – $y$) M $NaClO_4$ – $y$ M $K_2S_2O_8$ ($5 \cdot 10^{-3} \leqq y \leqq 0.02$; $I \simeq 0{,}1$) | 25 | 17 | – | EM | | [10] |

Unter Annahme der Gleichgewichte

$$[At(H_2O)_2]^+ \rightleftharpoons [At(H_2O)OH] + H^+ \; ; \quad K_s$$

$$[At(H_2O)_2]^+ + X^- \rightleftharpoons [At(H_2O)X] + H_2O \; ; \quad K_1$$

$$[At(H_2O)X] + X^- \rightleftharpoons AtX_2^- + H_2O \; ; \quad K_2$$

wurden die Gleichgewichtskonstanten (X = Cl, Br, I) aus den Elektromigrationsdaten ermittelt.

$$u_2 = \frac{u_{At(\theta)^+} + K_1 \cdot K_2 \cdot u_{AtCl^-} \cdot C_{Cl^-}^2}{1 + \dfrac{K_s}{C_{H^+}} + K_1 \cdot C_{Cl^-} + K_1 \cdot K_2 \cdot C_{Cl^-}^2}$$

und

$$u_m = \frac{u_{AtX_2^-} \cdot K_2 \cdot C_{X^-}}{1 + K_2 \cdot C_{X^-}}$$

für den Fall, daß $K_1 \gg K_2$ ist.

Die Ergebnisse sind in Tab. 2 angegeben. Sie stimmen gut mit den von *E. H. Appelmann* aus Extraktionsuntersuchungen ermittelten Daten [17] überein. In Tab. 2 sind weiter für den $AtI_2^-$-Komplex die Stabilitätskonstanten $K_2$ für verschiedene Temperaturen zusammengefaßt [10]. Hieraus ergeben sich für die freie Reaktionsenthalpie $\Delta_R G_2$, die Reaktionsenthalpie $\Delta_R H_2$ und die Reaktionsentropie $\Delta_R S_2$ die Werte:

$$\Delta_R G_2 \simeq 19 \text{ KJ/mol im } T\text{-Bereich } 291-313 \text{ K}$$

$$\Delta_R H_2 \simeq 26 \text{ KJ/mol}$$

$$\Delta_R S_2 \simeq 0{,}02 \text{ KJ/mol} \cdot \text{K}$$

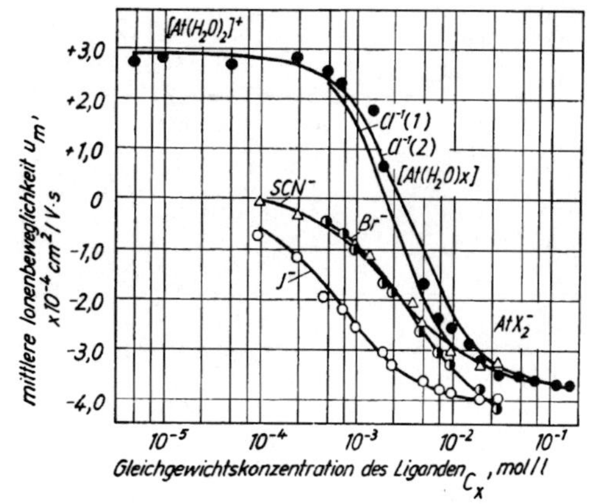


**Abb. 1.** Elektromigration des $At(0)^+$ in halogenid- und pseudohalogenidhaltigen Elektrolyten bei 298 K
Punkte – experimentelle Mittelwerte ($n \geqq 3$); Kurven – berechnet mit den Parametern in Tab. 2

Tab. 3. Ladungszahl $z$ und Koordinationszahl $n$ für $AtX_n^z$

| $X^-$ | $u_{At(0)^+}$ in $10^{-4}$ cm²/V·s | $u_{AtX_n^z}$ in $10^{-4}$ cm²/V·s | $z$ | $n$ |
|---|---|---|---|---|
| Cl⁻ | +3,70¹) | −3,80¹) | −1,03 | 1,97 |
| Br⁻ | +4,23²) | −4,70¹) | −1,11 | 1,90 |
| J⁻ | +4,23²) | −4,15¹) | −0,98 | 2,02 |
| SCN⁻ | +4,23²) | −3,7¹) | −0,88 | 2,14 |
| | | Mittelwert | ∼(−1) | ∼2 |

¹) s. Bemerkungen in Tab. 2; ²) s. Tab. 1

Eine Abschätzung der Ladung und der Ligandenzahl ist über das *Stokes*-Gesetz und die Gleichung von *Bjerrum* unter den in [23] angegebenen Bedingungen möglich:

$$Z = \frac{Z_{At(0)^+} \cdot u_{AtX_n^Z}}{u_{At(0)^+}};$$

$$n = \frac{u_{At(0)^+} - u_{AtX_n^Z}}{u_{At(0)^+}} \cdot \frac{Z_{At(0)^+}}{Z}$$

Unter Verwendung der in Abb. 1 und Tab. 1 und 2 angegebenen experimentellen Elektromigrationsdaten und mit $Z_{At(0)^+} = +1$ ergeben sich die in Tab. 3 zusammengefaßten Angaben für die Ladung und Zusammensetzung der Halogenidkomplexe des Atats.

Die Bildung von Astathalogenverbindungeng mit einem höheren Halogenanteil als 2 wurde für das $AtCl_2^- \cdot Cl_2$ auf der Grundlage von Adsorptionsuntersuchungen an Ionenaustauschern von *Chalkin* und Mitarbeitern [5] wahrscheinlich gemacht. Erste Elektromigrationsuntersuchungen zur Reaktion von $At^-$ mit elementarem Iod in wäßrigen Lösungen erbrachten den Nachweis, daß analog den Reaktionen des $I^-$ mit Iod [24] ein Ion gebildet wird, dessen Ionenbeweglichkeit erheblich geringer als die des $AtI_2^-$ ist. Die Bildung von $AtI_4^-$ ist wahrscheinlich [25].

### 2.2. *Pseudohalogenidkomplexe*

Die ersten Mitteilungen über die Bildung und Charakterisierung von Komplexverbindungen des $At(0)^+$ mit Rhodanid- und Cyanid-Ionen erfolgten 1982 [26, 27]. Für den Rhodanidkomplex des $At(0)^+$ konnte über die Bestimmung der Ionenbeweglichkeiten in freien Elektrolyten in Abhängigkeit von der $SCN^-$-Konzentration gezeigt werden [26], daß der Astatrhodanidkomplex analog den Halogenidkomplexen zusammengesetzt ist. Die Stabilitätskonstante für die Reaktion

$$At(SCN) + SCN^- \rightleftharpoons At(SCN)_2^-; \quad K_2$$

beträgt bei 298 K und einer Ionenstärke von $I = 0{,}05$ mol/l

$$K_2 = 400 \pm 30 \text{ l} \cdot \text{mol}^{-1}.$$

Hinsichtlich seiner Stabilität ordnet sich der Komplex zwischen dem $AtI_2^-$ und $AtBr_2^-$ ein (Tab. 2, Abb. 1). Der Komplex ist bis zu einem pH-Wert von $\leqq 8$ stabil. Die bisher vorliegenden Resultate zur Bildung von Astat-Cyanidverbindungen bestätigen deren Existenz in wäßrigen Lösungen. Eine ausführliche Charakterisierung steht noch aus. Erste Untersuchungsergebnisse zur Verbindungsbildung von $At(0)^+$ mit $C(CN)_3^-$ und $N_3^-$ liegen vor [28].

### 2.3. *Verbindungen des $At(0)^+$ mit $NO_3^-$, $SO_4^{2-}$, $Cr_2O_7^{2-}$, $S_2O_8^{2-}$*

Von *Chalkin* und Mitarbeitern [5, 18] wurde an Hand von Ionenaustauschuntersuchungen eine Verbindungsbildung des $At(0)^+$ mit $NO_3^-$, $SO_4^{2-}$ und $Cr_2O_7^{2-}$ nachgewiesen und die Gleichgewichtskonstanten abgeschätzt. Elektromigrationsuntersuchungen führten für das $Cr_2O_7^{2-}$ zu gleichen Aussagen und weisen auch auf eine Reaktion des $At(0)^+$ mit $S_2O_8^{2-}$ hin [10]. Weitere Angaben sind in Tab. 2 zusammengefaßt.

### 2.4. *Kationische Astat(I)-Komplexe mit schwefelhaltigen Liganden*

Im Resultat von Untersuchungen zur Extraktion anorganischer Formen des Astats in Gegenwart verschiedener Reagenzien wurde von *G. M. W. Visser* [27] erstmals die Verbindungsbildung des Astats mit schwefelhaltigen Liganden, wie Thioharnstoff (THS) und Thioacetamid (TAA) beschrieben. Analog den Iodverbindungen [29, 30] wird eine Zusammensetzung $At(THS)_2^+$ und $At(TAA)_2^+$ angenommen.

Elektromigrationsstudien zur Synthese und Charakterisierung dieser Gruppe von Astat(I)-Komplexverbindungen in wäßrigen Lösungen bestätigten die Bildung der Astatthioharnstoffkomplexe. Für Thioacetamid konnte unter gleichen Bedingungen keine Verbindungsbildung nachgewiesen werden [31]. In der gleichen Arbeit wurde weiter gezeigt, daß $At(0)^+$ mit den verschiedenen substituierten Methylthioharnstoffen sowie mit Pyrrolidino-

Tab. 4. Elektromigration von Astat(I)-Thioharnstoff-Kationen $At(L)_n^+$ in verschiedenen Elektrolyten bei 298 K (nach [31])

| Ligand (L) | Elektrolyt | $u_m$ in $10^{-4}$ cm²/V·s |
|---|---|---|
| Thioharnstoff | 0,02 M NaClO₄ | 2,62 |
| | 0,02 M NaClO₄ + 0,02 M L | 2,62 |
| | 0,02 M HClO₄ + 0,02 M L | 2,60 |
| | 0,02 M HCl + 0,02 M L | 2,60 |
| | 0,02 M HBr + 0,02 M L | 2,23 |
| | 0,02 M HJ + 0,02 M L | 1,50 |
| Trimethylthioharnstoff | 0,02 NaClO₄ | 1,75 |
| | 0,02 M NaClO₄ + 0,02 M L | 1,75 |
| | 0,02 M HBr + 0,02 M L | 1,70 |
| Tetramethylthioharnstoff | 0,02 M NaClO₄ | 1,80 |
| | 0,02 M NaClO₄ + 0,02 M L | 1,80 |
| | 0,02 M HBr + 0,02 M L | 1,66 |
| Pyrolidinothioharnstoff | 0,02 M NaClO₄ | 1,55 |
| | 0,02 M NaClO₄ + 0,02 M L | 1,55 |
| | 0,02 M HBr + 0,02 M L | 1,70 |

thioharnstoff in wäßrigen Lösungen stabile kationische Astatkomplexe bildet. Als Ausgangsverbindungen zur Synthese können $At(0)^+$ und Astathalogenidkomplexe, insbesondere $AtCl_2^-$ eingesetzt werden. In Tab. 4 sind einige Angaben zu den Ionenbeweglichkeiten von $At(0)^+$-Komplexen mit schwefelhaltigen Liganden zusammengefaßt.

Weiterführende Untersuchungen ergaben [32, 33], daß neben weiteren Thioharnstoffderivaten auch N-Acylthioharnstoffe kationische Astatverbindungen bilden.

Elektromigrationsuntersuchungen in HCl-haltigen Elektrolyten, in welchen ohne Gegenwart eines weiteren Komplexbildners $AtCl_2^-$ gebildet wird, erlauben es, gesicherte Aussagen zur Verbindungsbildung des $At(0)^+$ auch beim Vorliegen von Neutralkomplexen zu erhalten. Die Verwendung von gemischten Elektrolyten, die neben dem schwefelhaltigen Komplexbildner noch Halogenidionen bzw. $SCN^-$ enthalten, erlaubten erste Abschätzungen zur Stabilität dieser neuen Verbindungsgruppe von kationischen Astatverbindungen (s. Abb. 2). In Tab. 5 sind weitere Liganden zusammengefaßt, die außer den bereits genannten Thioharnstoffderivaten mit $At(0)^+$ zu kationischen Komplexverbindungen reagieren.

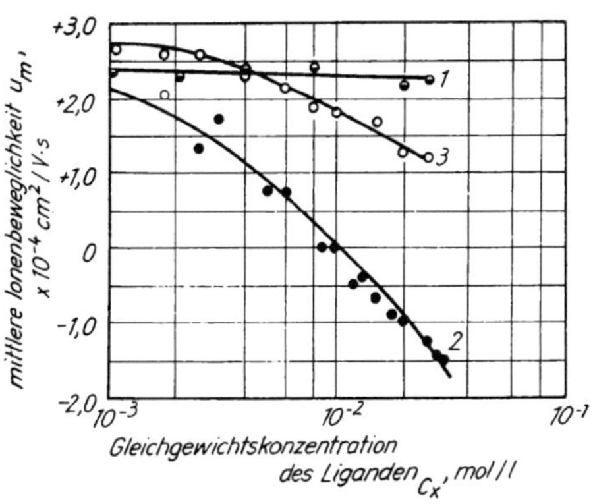


Abb. 2. Elektromigration von $At(0)^+$ bzw. $AtX_2^-$ bei 298 K in 0,02 M $HClO_4$ – $(0{,}03 - y)$ M $NaClO_4$ – $y$ M (Na/K)X – $z$M Thioharnstoff ($1 \cdot 10^{-3} \leq y \leq 0{,}03$)
(1) $X = Cl^-$, $z = 2{,}5 \cdot 10^{-3}$; (2) $X = I^-$, $z = 2{,}5 \cdot 10^{-3}$; (3) $X = SCN^-$, $z = 1{,}6 \cdot 10^{-3}$
(nach [32, 33])

Aus den vorliegenden Ergebnissen können folgende Aussagen abgeleitet werden:

– Die Ionenbeweglichkeit von $2{,}6 \cdot 10^{-4}$ cm² s⁻¹ v⁻¹ bei 298 K, $I = 0{,}02$ mol · l⁻¹ für den Astatthioharnstoffkomplex weist auf eine Ladung von +1 hin.
– Die Ionenbeweglichkeiten für die Verbindungen des $At(0)^+$ mit Thioharnstoffderivaten sowie weiteren schwefelhaltigen Liganden nehmen mit zunehmender Ligandengröße ab. Für N-Acylthioharnstoff-Astat-Komplexe liegen sie bei 298 K in 0,02 M $HClO_4$ im Bereich von $1 \cdot 10^{-4}$ cm² s⁻¹ v⁻¹.
– Die Zusammensetzung der kationischen Astatverbindungen ist analog der des $[At(H_2O)_2]^+$, der Halogeni- und Pseudohalogenidkomplexe $AtX_2^-$ ($X = Cl^-$, $Br^-$, $I^-$, $SCN^-$, $C(CN)_3^-$) zu $[At(Y)_2]^+$ (Y = THS, THS-Derivate) anzunehmen.
– Die Ionenbeweglichkeiten sind unter den angegebenen Bedingungen [32] unabhängig von der Ligandenkonzentration im Elektrolyten. Die $[At(Y)_2]^+$-Komplexe sind demnach wesentlich stabiler als die bisher bekannte stabilste Astatkomplexverbindung $AtI_2^-$.

## 2.5. *Hydrolyseverhalten von At(I)-Verbindungen*

*E. H. Appelmann* [17] beobachtete bei der Extraktion von Astat aus sauren iodhaltigen Lösungen mit $CCl_4$ im pH-Bereich 1–5 die Unabhängigkeit des Verteilungskoeffizienten von der Acidität. Beim pH ≃ 7 und bei einer Iodkonzentration von $10^{-4}$ mol · l⁻¹ sank er merklich ab. Die Hydrolysekonstante für die Reaktion

$$AtI + H_2O \rightleftharpoons HOAt + I^- + H^+; \qquad K_H$$

wurde hieraus zu $K_H \lesssim 10^{-11}$ mol · l⁻¹ abgeschätzt, für die analoge Reaktion des IBr beträgt

$$K_H \simeq 1{,}7 \cdot 10^{-7}\ \text{mol} \cdot \text{l}^{-1}.$$

Ergebnisse zur Hydrolyse des $At(0)^+$, der Halogenid- und Pseudohalogenidkomplexe und der kationischen Astatthioharnstoffverbindungen liegen nur in begrenztem Umfang vor. Sie beschränken sich auf Aussagen zur Beständigkeit der Komplexe bei unterschiedlichen ph-Werten und auf die Charakterisierung des Endproduktes [2, 10, 20, 26].

In Abb. 3 sind Ergebnisse zur Ionenbeweglichkeit von $AtBr_2^-$, $At(SCN)_2^-$, $AtI_2^-$ bei konstanter Ionenstärke und variablem pH zusammengefaßt [10, 26].

Es besteht offensichtlich ein Zusammenhang zwischen der Stabilitätskonstante $K_2$ der Komplexe und dem Umwandlungsbereich. Der $AtI_2^-$-Komplex als stabilste Ver-

Tab. 5. Elektromigration kationischer Komplexe des At(I) mit S-(O)-Donatoren in perchlor- bzw. salzsauren Elektrolyten bei 298 K (nach [32, 33])

| Ligand | Zusammensetzung des Elektrolyten | | | | Ionenbeweglichkeit |
|---|---|---|---|---|---|
| | $c_{Ligand}$ in $10^{-3}$ mol · l⁻¹ | $c_{HClO_4}$ in $10^{-2}$ mol · l⁻¹ | $c_{NaClO_4}$ in $10^{-2}$ mol · l⁻¹ | $c_{HCl}$ in $10^{-2}$ mol · l⁻¹ | $u_{Kation}$ in $10^{-4}$ cm²/V · s |
| 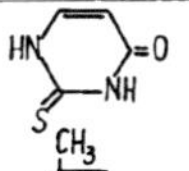 | gesättigt<br>gesättigt | 2,0<br>– | 3,0<br>3,0 | –<br>2,0 | + (2,40 ± 0,06)<br>+ (0,70 ± 0,02) |
| 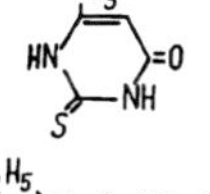 | gesätitgt<br>gesättigt | 2,0<br>– | 3,0<br>3,0 | –<br>2,0 | + (2,45 ± 0,12)<br>+ (2,42 ± 0,08) |
| 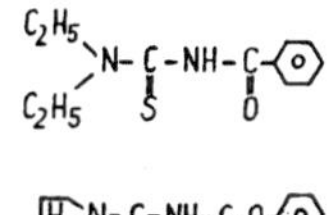 | 2,5<br>2,5 | 2,0<br>– | 3,0<br>3,0 | –<br>2,0 | + (1,38 ± 0,23)<br>+ (0,80 ± 0,12) |
| 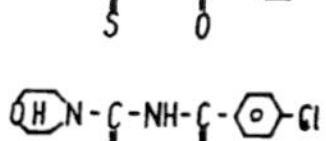 | 2,5<br>2,5 | 2,0<br>– | 3,0<br>3,0 | –<br>2,0 | + (1,15 ± 0,29)<br>+ (1,00 ± 0,21) |
| 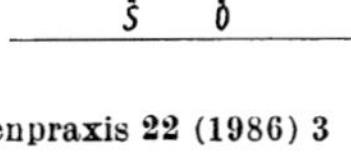 | 2,5<br>2,5 | 2,0<br>– | 3,0<br>3,0 | –<br>2,0 | + (0,76 ± 0,21)<br>+ (0,83 ± 0,19) |

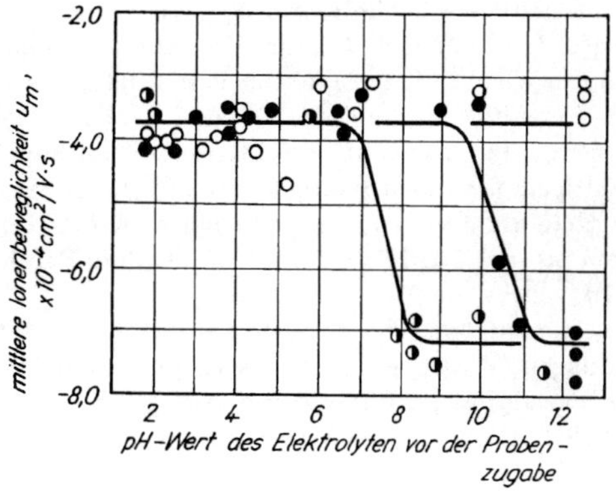


Abb. 3. Elektromigration von $AtX_2^-$ bei 298 K in 0,03 M NaX − (0,02 − *y*) M $NaClO_4$ − *y* M $HClO_4$ oder NaOH ($0 \leq y \leq 0{,}02$) $X^-$ − $Br^-$ (●); $I^-$ (○); $SCN^-$ (◑) (nach [20, 26])

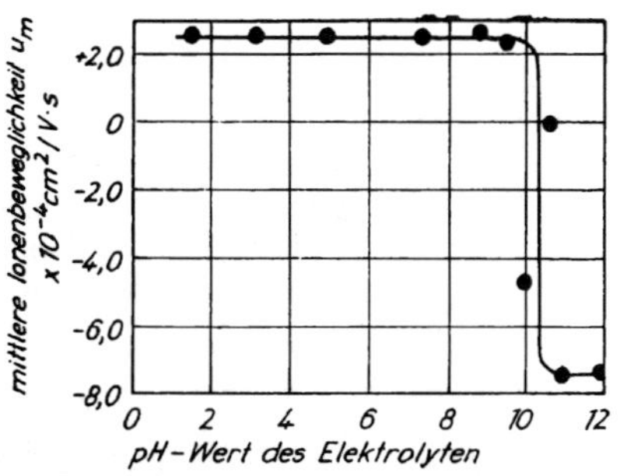


Abb. 4. Elektromigration von $At(Thioharnstoff)_2^+$ bei 298 K in 0,02 M Thioharnstoff − (0,02 − *y*) M $NaClO_4$ − *y* M $HClO_4$ oder NaOH ($0 \leq y \leq 0{,}02$) (nach [32])

bindung dieser Gruppe ist bis zu einem pH $\lesssim 12$ stabil, während $AtBr_2^-$, $At(SCN)_2^-$ und auch das $AtCl_2^-$ bereits bei niedrigeren pH-Werten Hydrolysereaktionen zeigen. Charakteristisch ist, daß für alle Verbindungen als Endprodukt ein Anion resultiert, dessen Ionenbeweglichkeit bei 298 K der des $At^-$ unter gleichen Bedingungen entspricht.

Gleiches gilt für die Hydrolysereaktionen des $[At(H_2O)_2]^+$. Bei Ionenstärken von $1 \cdot 10^{-2} - 1 \cdot 10^{-1}$ mol · $l^{-1}$ und 298 K bilden sich im pH-Bereich $\gtrsim 7$ Anionen mit Ionenbeweglichkeiten von $7{,}0 - 7{,}4 \cdot 10^{-4}$ cm²/V·s. Unter gleichen Bedingungen durchgeführte Paralleluntersuchungen mit $At^-$ führten zu identischen Ionenbeweglichkeiten [10].

Erste Untersuchungen zum Hydrolyseverhalten des $[At(THS)_2]^+$ [32] ergaben ein prinzipiell gleiches Bild. Der Astatthioharnstoff-Komplex ist bei 298 K bis zu einem pH-Bereich von 10–11 stabil. Bei einer weiteren Zunahme des pH-Wertes wird auch in diesem Falle ein Hydrolyseprodukt gebildet, dessen Ionenbeweglichkeit der des $At^-$ entspricht (s. Abb. 4).

Von *G. M. W. Visser* wird auf der Grundlage von Extraktionsuntersuchungen ein Reaktionsschema diskutiert, welches in Abwesenheit von Reduktionsmitteln $At(OH)_2^-$ als Endprodukt der Hydrolyse postuliert [27]. Zur Charakterisierung des offensichtlich gleichen Hydrolyseproduktes von $At(0)^+$ und der Astat(I)-Komplexverbindungen sind vom gegenwärtigen Kenntnisstand der Astatverbindungen ausgehend, folgende Faktoren zugrunde zu legen:

- Außer dem $OH^-$-Ion gibt es keine Anionen mit einer Ladung von −1, deren Ionenbeweglichkeiten im Bereich der Halogenidionen liegen.
- Anionen des Astats vom Typ $AtO_n^-$ ($n = 2, 3, 4$) besitzen Ionenbeweglichkeiten zwischen $2{,}5 - 5 \cdot 10^{-4}$ cm²/V·s. Sie ordnen sich in die Reihe der analogen Halogenverbindungen ein. $At(OH)_2^-$ sollte sich ebenfalls in diesen Bereich einordnen.
- Die Ionenbeweglichkeit des $[At(H_2O_2]^+$ ist mit der der Oxianionen vergleichbar, gleiches gilt für die Halogenid- und Pseudohalogenid-Komplexe des Astats.

Daraus folgt: Reaktionsabläufe, die eine Kombination von Hydrolyse und Reduktionsreaktionen sowie als Zwischenprodukt $At(OH)_2^-$ und Endprodukt $At^-$ enthalten, entsprechen bisher am besten den experimentellen Befunden.

## Schlußfolgerungen

Astat zeigt in den Oxydationsstufen −1, +3, +5 und +7 die typischen Eigenschaften der Halogene. Im Oxidationszustand +1 besitzt es weiterhin „metallische" Eigenschaften, die sich mit denen der Übergangsmetalle Cu(I), Ag(I) und Au(I) vergleichen lassen, wobei die Ähnlichkeit im Reaktionsverhalten mit dem des Ag(I) besonders augenscheinlich ist. Ordnet man das $At(0)^+$ in das Modell der harten und weichen Säuren von Pearson ein [27], so besitzt es eine Mittelstellung zwischen harter und weicher Säure, wobei der „weiche" Charakter überwiegt.

Der bisher erreichte Erkenntnisstand zur Chemie des Astats ist bei weitem nicht ausreichend, um eine vollständige Übersicht über die chemischen Reaktionen des At(I) und das Verhalten seiner Verbindungen zu ermöglichen.

Der durch die Anwendung der Elektromigrationsmethode erreichte Erkenntniszuwachs zur Synthese und zum Verhalten von At(I)-Verbindungen in wäßrigen Lösungen bildet jedoch eine gute Grundlage für weiterführende Untersuchungen. Der Einsatz von gemischten wäßrig-nichtwäßrigen Elektrolyten sowie von wasserfreien Lösungsmitteln erlaubt es, Komplexbildner in die Untersuchungen einzubeziehen, deren Wasserlöslichkeit gering ist. Neben der Erweiterung der Kenntnisse zur Chemie des 5. Halogens ist die Anwendung bereits charakterisierter, mit $^{211}$At-markierten Verbindungen für die Nuklearmedizin von besonderem Interesse.

Eingegangen am 26. 4. 1985
Überarbeitete Fassung am 28. 5. 1985

## Literatur

[1] *G. L. Johnson, R. F. Leininger, E. Segré*, J. Chem. Phys. **17** (1949) 1
[2] *I. Dreyer, R. Dreyer, V. A. Chalkin*, Radiochem. Radioanal. Letters **36** (1978) 389
[3] *Do Kim Tjung, I. W. Dudowa, V. A. Chalkin*, Radiochimija **15** (1973) 548
[4] *Wan Fu-Dsun, Ju. W. Norsejew, V. A. Chalkin, Tschao Toa Nan*, Radiochimija 5 (1963) 351
[5] *V. A. Chalkin, E. Herrmann*, Isotopenpraxis **11** (1975) 333
[6] *V. Jedinakova, I. Celeda*, Coll. Czech. Chem. Comm. **36** (1971) 3071
[7] *R. P. Bell, E. Gelles*, J. Chem. Soc. **1951**, 2737
[8] *De Puz M.*, J. Chem. Phys. **52** (1970) 3362
[9] *M. Milanov, V. Doberenz, V. A. Khalkin, A. Marinov, J. Radioana* Nucl. Chem., Articles **82**/2 (1984) 291
[10] *F. Rösch*, Dissertation A, TU Dresden 1984
[11] *K. H. Mischenko, H. E. Flis*, J. Angew. Chem. (USSR) **30** (1957) 665
[12] *L. Vasaros, Ju. V. Norsejew, D. D. Nhan, V. A. Chalkin*, Radiochem. Radioanal. Letters **54** (1982) 239
[13] *E. H. Appelman, E. N. Sloth, M. N. Studier*, Inorg. Chem. **5** (1966) 766

[14] *J. Merinis, Y. Legoux, G. Boussieres*, Radiochem. Radioanal. Letters **11** (1972) 59
[15] *G.-J. Meyer*, Ber. Kernforschungsanlage Jülich 1418 (1977)
[16] *G.-J. Meyer, K. Rössler*, Radiochem. Radioanal. Letters **25** (1976) 377
[17] *E. H. Appelman*, Dissertation Univ. Kalifornien, Berkeley 1960 (URCL-9025)
[18] *V. A. Chalkin, E. Herrmann, Ju. W. Norsejew, I. Dreyer*, Chemikerzeitung **101** (1977) 470
[19] *I. Dreyer, R. Dreyer, V. A. Chalkin, M. Milanov*, Radiochem. Radioanal. Letters **403** (1979) 145
[20] *R. Dreyer, I. Dreyer, F. Rösch, G.-J. Beyer*, Radioanal. Radiochem. Letters **54/3** (1982) 165
[21] *M. Milanov, W. Doberenz, V. A. Chalkin*, in Vorbereitung
[22] *R. Dreyer, I. Dreyer, F. Rösch*, Z. für Chemie **22** (1982) 54
[23] *A. V. Stepanov, E. K. Kortcemnaja*, Die Elektromigrationsmethode in der anorganischen Analyse. Moskau: Verlag Chemie 1979 (*russ.*)
[24] *F. Kilz, W. Kühnel*, Dissertation A, TU Dresden 1982
[25] *R. Dreyer, S. Fischer, I. Dreyer*, Radioanal. Chem., in Vorbereitung
[26] *R. Dreyer, I. Dreyer, M. Pfeiffer, F. Rösch*, Radiochem. Radioanal. Letters **554** (1982) 207
[27] *G. W. M. Visser*, Dissertation, Univ. Amsterdam 1982
[28] *S. Fischer, R. Dreyer*, Radioanal. Chem., in Vorbereitung
[29] *G. H. Y. Lan, H. Hope*, Acta Crystallogr. (Kopenhagen) **B28** (1972) 643
[30] *R. P. Lang*, J. Am. Chem. Soc. 84 (1982) 1185
[31] *R. Dreyer, I. Dreyer, F. Rösch, S. Fischer*, Z. für Chemie **23/0** (1983) 346
[32] *R. Dreyer, I. Dreyer, S. Fischer, H. Hartmann, F. Rösch*, Konferenzbericht Hauptjahrestagung der Chem. Ges. der DDR, Dez. 1984 Dresden
[33] *R. Dreyer, I. Dreyer, S. Fischer, H. Hartmann, F. Rösch*, Radioanal. Chem., in Vorbereitung

Isotopenpraxis 22 (1986) 3, S. 87 bis 89

# Characterization of Chromized Metallic Surfaces by Means of Radioactive $^{51}$Cr

*V. Rößiger, A. Freyer, E. Hartmann, C. Treutler*

(Central Institute of Isotope and Radiation Research Leipzig)[1]

*V. Brabec, O. Dragoun, A. Kovalík*

(Nuclear Physics Institute of the Czechoslovak Academy of Sciences, Řež)[2]

*The spatial distribution of Cr deposited on metallic surfaces at concentrations of about $10^{16}$ at/cm$^2$ was examined by detecting the radiation components emitted in the $^{51}$Cr decay. The autoradiography revealed a non-homogeneous Cr covering. Combined Auger electron and X-ray spectroscopies yielded information on the Cr concentration, especially in the 2 nm thick surface layer. This concentration was found to depend strongly on the chromatation procedure applied. The influence of multiple vacancies on the LMM Auger spectrum of vanadium was investigated theoretically.*

*Die räumliche Verteilung von Cr an metallischen Oberflächen im Konzentrationsbereich von ungefähr $10^{16}$ at/cm$^2$ wurde durch die emittierten Strahlenkomponenten des $^{51}$Cr-Zerfalls bestimmt. Die Autoradiographie ergab eine nicht-homogene Cr-Bedeckung. Die Auger-Elektronen- und Röntgenspektroskopie lieferte Informationen über die Cr-Konzentrationen, insbesondere in den äußeren 2 nm. Diese Konzentrationsverteilung war streng vom angewendeten Chromatierungsverfahren abhängig. Der Einfluß der Mehrfachvakanzen auf das LMM Auger-Spektrum des V wurde theoretisch untersucht.*

**Keywords**

Auger electron spectroscopy; autoradiography; chromium 51; spatial distribution, X-ray emission analysis

## 1. Introduction

Galvanic coatings or other technical surfaces are often covered by a small amount of Cr to inhibit their corrosion. In order to optimize the chromatation procedure one needs some information on depth and lateral distribution of the Cr atoms as well as on their chemical state. For typical technical samples under consideration recent PIXE measurements [1] yielded a total Cr concentration of about $10^{16}$ at/cm$^2$. This low concentration prevents or makes difficult the application of the usual methods of surface analysis such as electron microprobe for the investigation of lateral distributions and electron spectroscopies (XPS or AES) for chemical analysis or even depth profiling [2].

Addresses: [1]) DDR-7050 Leipzig, Permoserstr. 15; [2]) 250 68 Řež near Prague, ČSSR

In this work, we attempted to overcome the difficulties mentioned by measuring soft X-rays and *Auger* electrons emitted in the electron capture decay of $^{51}$Cr ($T_{1/2} = 27.7$ d).

## 2. Sample Preparation

Due to radiation protection it was not reasonable to label original industrial samples with $^{51}$Cr. Instead, laboratory specimens representative for certain technologies were prepared using the same basic materials as in industry. The PIXE measurements proved that the surface elemental composition of these samples corresponds to the industrial ones.

The necessary solution of radioactive $K_2CrO_4$ was made by oxidization of neutron irradiated $Cr_2O_3$ (specific activity of 0.8 GBq/mg, supplied by ZfK Rossendorf, GDR)

with $KBrO_3$ in aqueous solution. As a reference sample, a layer of metallic Cr was electrodeposited and annealed in vacuum according to [3].

## 3. Autoradiography

The Cr lateral distribution was investigated by means of autoradiography. A typical example is shown in Fig. 1 which was obtained by direct contact between film and sample. The black spots are Cr enrichments demonstrating a non-homogeneous covering. Although the lateral resolution was limited by the roughness of the technical samples, the autoradiographical images give important hints for further microanalytical studies.

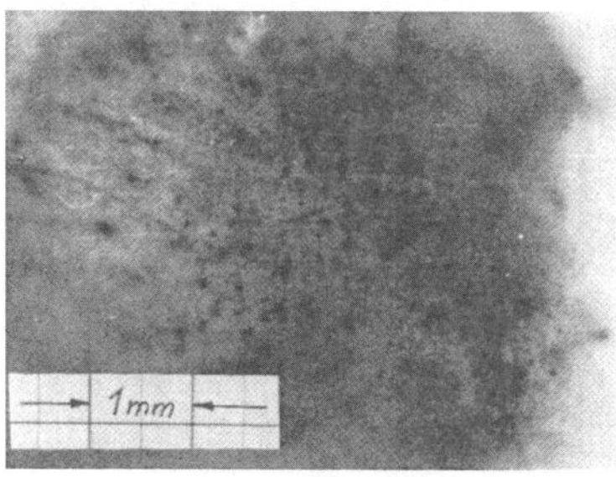


Fig. 1. Autoradiography of a chromized sample labeled with $^{51}Cr$ (sample No. 3 in Tab. 2)

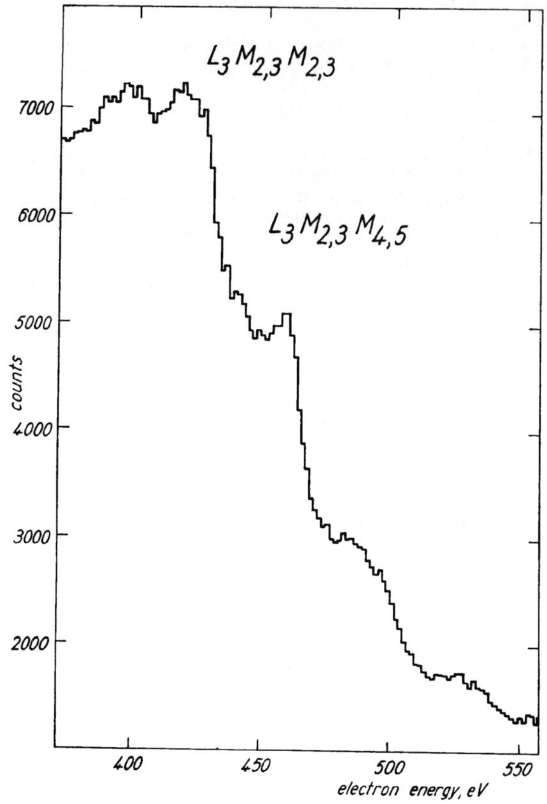


Fig. 2. Spectrum of the LMM *Auger* electrons of vanadium populated in the electron capture decay of $^{51}Cr$

## 4. Auger Electron Spectroscopy

### 4.1. *Shape of the LMM Spectrum*

The spectrum of the LMM *Auger* electrons of vanadium was measured e.g. by *Yin* et al. [4] who bombarded clean V metal with Mg $K_\alpha$ X-rays. The $L_3M_{2,3}M_{4,5}$ and $L_3M_{4,5}M_{4,5}$ lines recorded with 0·4 eV instrumental resolution exhibited a FWHM of 6·5 eV. Therefore, these lines could supply, at least in principle, some information on the chemical state of the investigated atoms [5].

The LMM *Auger* spectrum of V populated in the electron capture decay of $^{51}Cr$ was not, to our knowledge, yet reported. We carried out the measurements of such electron spectra using the electrostatic spectrometer described in [6]. Although we used radioactive samples of 8 mm in diameter we could not employ a better instrumental resolution than about 6 eV. This was due to the rather low portion of *Auger* electrons emerging from our technical samples without energy losses. An example of a spectrum is shown in Fig. 2.

There is some similarity among the *Auger* spectra obtained in this work and the spectra published for the 3 d transitions metals [7]. One should keep in mind, however, the difference between external and internal production of the inner shell vacancies. In the case of excitation with X-rays or electrons (both of not too high energy) single L vacancies are the starting point of the LMM transitions under consideration. In constrast to this situation, after electron capture excitation one is faced with a wide variety of multiple vacancies occuring in the initial state of the V LMM transition. This is a consequence of the pre-processes which transfer the vacancies from the K to the L shell. Due to the small fluorescence yield ($\omega_K = 0{\cdot}25$ for V) the *Auger* processes predominate and therefore the vacancy transfer is in most cases accompanied by a creation of additional vacancies in the outer (L and M) shells. The vacancy multiplication and its effect on the V LMM spectrum was simulated by the aid of a slightly modified computer code [8] for atomic Xα calculations. The results (Tab. 1) may be used for a qualitative interpretation of the measured spectra (Fig. 2) although a high background of inelastically scattered electrons did not enable us to compare details with the theoretical prediction.

### 4.2. *Cr Suface Concentration*

Since the mean free path length of 0·5 keV electrons in solids is about 1 nm [9] the intensity of the $L_3M_{2,3}M_{4,5}$ *Auger* line (Fig. 2) can be taken as a measure of the $^{51}Cr$ concentration within the surface layer of about 2 nm thickness. The "mean free path" of V K X-rays is about three orders of magnitude larger and therefore, the X-ray intensity is proportional to the total amount of $^{51}Cr$ atoms in the sample. We determined the absolute Cr concentrations by means of a calibrated X-ray detector taking into account the specific activity of the applied substance. (The same information could be obtained by means of the detection of the 320 keV γ-rays also emitted in the $^{51}Cr$ decay).

The experimental results are summarized in Tab. 2. Comparing three different chromizing procedures we noted considerable differences in the LMM *Auger* intensity which where not correlated with the total Cr concentrations. Thus a doubling of the exposure time of procedure I gave a double total content of Cr while the surface concentration remained the same (samples 1 and 2). The procedure II led to a higher Cr surface concentration at substantially lower total Cr amount. The procedure III provided both concentrations lower than in the previous cases but the

Tab. 1. Influence of additional vacancies created after $^{51}Cr$ decay on energies and intensities of the LMM *Auger* electrons in vanadium

| Pre-processes | Number of additional vacancies | % of all LMM transitions | Calculated energy shifts, eV | |
|---|---|---|---|---|
| | | | $L_{2,3}M_{2,3}M_{2,3}$ | $L_{2,3}M_{2,3}M_{4,5}$ |
| EC(K shell) + K X-ray EC (L shell) | 0 | 20 | 0 | 0 |
| EC(K shell) + KLL *Auger* | 1 (L) | 38 | 7 ($L_1$) 12 ($L_{2,3}$) | 9 ($L_1$) 14 ($L_{2,3}$) |
| EC(K shell) + KLM *Auger* | 1 (M) | 4 | −15 ($M_{2,3}$) | −14 ($M_{2,3}$) |
| EC(K shell) + KLL *Auger* + LMM *Auger* | 2 (M) | 38 | −31 (2 $M_{4,5}$) | −30 (2 $M_{4,5}$) |

The energy shifts were calculated for additional vacancies specified in brackets.

Tab. 2. Total and surface Cr concentrations in samples prepared by various chromizing procedures

| Sample No. | Chromizing procedure | Chromizing time, s | Total Cr concentration $10^{16}$ at/cm$^2$ | Surface Cr concentration relat. units |
|---|---|---|---|---|
| 1 | I | 15 | 1·6 ± 0·2 | 0·5 ± 0·1 |
| 2 | I | 30 | 3·2 ± 0·3 | 0·5 ± 0·1 |
| 3 | II | 20 | 0·7 ± 0·1 | 0·6 ± 0·1 |
| 4 | III | 20 | 0·5 ± 0·1 | 0·3 ± 0·1 |

relative surface concentration was higher than for procedure I.

The intensity of the KLL or KLM *Auger* electrons not measured in this work would supply information on the Cr concentration in a surface layer of about 15 nm thickness.

## Acknowledgement

The authors are indebted to Professor Dr. *G. Brunner* for stimulating and helpful discussions.

Received April 9, 1985

## References

[1] *W. Frank, M. Lenk, V. Rößiger,* Isotopenpraxis **20** (1984) 227; *V. Rössiger,* Nucl. Instrum. Meth. **196** (1982) 483
[2] *M. Pijolat, G. Hollinger,* Surface Sci. **105** (1981) 114
[3] *M. Mehlhorn, R. G. Albridge,* Z. Phys. **175** (1963) 506
[4] *L. I. Yin, T. Tsang, I. Adler,* Phys. Rev. **B15** (1977) 2974
[5] *C. D. Wagner,* J. Electron Spectrosc. **10** (1975) 305
[6] *D. Varga, I. Kádár, A. Kövér, I. Czerny, G. Mórik, V. Brabec, O. Dragoun, A. Kovalik, J. Adam,* Nucl. Instrum. Meth. **192** (1982) 277
[7] *J. C. Fuggle,* in: Electron Spectroscopy: Theory, Techniques and Applications, Vol. 4, p. 85. New York: Academic Press 1981
[8] *F. Herman, S. Skillman,* Atomic Structure Calculations. Englewood Cliffs: Prentice Hall 1963
[9] *G. Oertel, E. H. Weber,* phys. stat. sol. (a) **43** (1977) 141

Isotopenpraxis **22** (1986) 3, S. 89

## Buchbesprechung

### Die Mikroanalyse der Leistungsfähigkeit von Rechnersystemen

*B. Beizer. Aus dem Engl. von V. V. Martynjuka Moskau: Verlag Radio i swjas 1983. 360 S., div. Abb. (in Russisch); geb. Rbl. 2,– ; M 10,–*

Das Buch ist eine Übersetzung des 1978 bei Litton Educational Publishing, Inc. herausgegebenen Buches „Microanalysis of computer system performance". Gegenwärtig steht in Zusammenhang mit vielfältiger Rechentechnik, was die Prozessoren als auch die Peripherie anbetrifft, vor den potentiellen Nutzern dieser Technik die Aufgabe einer qualifizierten Bewertung der Leistungsfähigkeit der Rechentechnik. Dieses Problem wird für Mikro- und Minirechentechnik sehr ausführlich vom Autor sowohl für Hard- als auch für Software behandelt. Im vorliegenden Buch werden vielfältige Methoden zur Untersuchung der Leistungsfähigkeit von Rechnersystemen dargestellt. Dabei wird auf die Strukturen, die Wirkungsprinzipien und die Organisation von Mikrorechnern eingegangen. Der Autor benutzt analytische Methoden. Dieses Herangehen wird von ihm als Mikroanalyse bezeichnet. Bei dieser Methode beginnt die Untersuchung auf dem niedrigsten Niveau des Systems, auf dem Niveau des Kodes und endet mit dem Aufstellen von Modellen. Demgegenüber wird bei der Makroanalyse, die nicht betrachtet wird, die Systemcharakteristik auf das Grundverhalten des Gesamtsystems zurückgeführt. Charakteristisch ist für die Makroanalyse die Messung und Modellierung.

Die Mikroanalyse ist wegen des großen Umfangs analytischer Berechnungen auf die Analyse und Optimierung vergleichsweise einfacher Systeme beschränkt und nicht geeignet zur Analyse kommerzieller Operationssysteme.

In 16 Kapiteln werden ausgehend von Strukturen, Ressourcen und Anwendungen der Rechnersysteme, nachdem Parameter und Maß der Leistungsfähigkeit definiert worden sind, die statistische Analyse dargelegt, Grenzen und Möglichkeiten der analytischen Methoden diskutiert, Warteschlangentheorie, zyklische und asynchrone Systeme behandelt und schließlich die Tauglichkeit der Modelle geprüft. Der Autor versucht den Stoff durch humoristische Fragestellungen aufzulockern.

Das Buch ist für einen breiten Kreis von Spezialisten geschrieben, die sich mit dem Aufbau und der Nutzung von Mikrorechnern in automatisierten Systemen beschäftigen. Die Anforderungen an den Leser sind relativ hoch. Es wird beim Studium die ständige Mitarbeit verlangt. Dazu dienen auch Übungsfragen und -aufgaben am Ende jedes Abschnitts. Es werden Kenntnisse auf dem Gebiet der Mikrorechentechnik, der Programmierung (Assembler) und Mathematik (Algebra und Statistik) vorausgesetzt.

*F. Riedel*

Isotopenpraxis 22 (1986) 3, S. 90 bis 93

# Further Studies on Gamma Radiolysis of Some Mono- and Disaccharides in the Solid State

*M. M. Abou-Sekkina**), *M. A. Sakran*, *A. A. Saafan*

(Chem. Dept., Faculty of Science, Tanta University, Tanta, Egypt)

*Glucose and Lactose molecules have been prepared and recrystallized from bidistilled water. Extensive measurements have been carried out on these materials in the solid state comprising infrared; electronic absorption spectra before and after exposure to an absorbed energetic γ-ray dose. These materials were chosen as representatives of different types of sugars including mono- and disaccharides. They have hydrogen bonding of variant origins. The relative strength of the cyclic structures and the magnitude of γ-radiation damage were evaluated in each case. Gamma radiotic oxidation mechanism was established for both mono- and disaccharides investigated. A correlation between γ-radiation damage radiotic and relative strength of cyclic structures was put forward and established for the first time.*

*Es wurden Glukose- und Laktose-Moleküle gewonnen und aus bidestilliertem Wasser rekristallisiert. Mit diesen Stoffen wurden umfangreiche Messungen einschließlich Infrarotmessungen im festen Zustand durchgeführt und Elektronen-Absorptionsspektra vor und nach der Bestrahlung mit einer energiereichen Gammastrahlung ermittelt. Sie wurden als Muster verschiedener Zuckerarten einschließlich Mono- und Disacchariden gewählt. Sie weisen Wasserstoffverbindungen verschiedener Herkunft auf. Die relative Stärke dieser zyklischen Strukturen und der Umfang der Gamma-Strahlenschäden wurden in jedem Falle ausgewertet. Der Mechanismus der Gamma-Radiooxidation bei Mono- und Disacchariden wurde untersucht. Eine Beziehung zwischen den Gamma-Strahlenschäden und der relativen Stärke der zyklischen Struktur wurde zum ersten Mal hergestellt.*

## Keywords

disaccharides; electron spectra; gamma radiation; infrared spectra; monosaccharides; radiolysis

## Introduction

In their paper *Hassel* and *Ottar* [1] suggested and introduced conformational analysis that the facts established for the cyclohexane system could also be applied to sugars in their pyranose forms. The pioneering work on the conformations of sugars was carried out by *Reeves* [2]. He proved that the pyranoid sugars exist in chair conformations. Glucose exists as a cyclic (hemiacetal) form; and it was soon found two (α and β) crystalline forms of glucose which are:

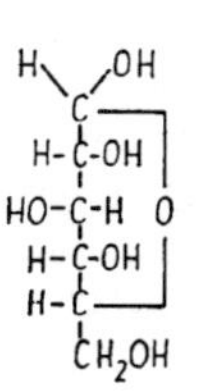

α-D-glucopyranose

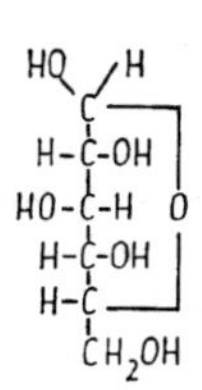

β-D-glucopyranose

*Isbell* [3] suggested the use of $C_1$ for the stable form of glucose in both the D and L series and $C_2$ for the other chair form. Conformational analysis of the pyranoid sugars is based on the assumption that the geometry of the pyranose ring is the same as that of cyclohexane, careful study of glucopyranosides, galactopyranosides convinced *Reeves* [2] that the boat forms play no significant role in the equilibrium of sugar conformations, on the assumption that only chair forms need to be considered.

*) Author for correspondence to be addressed

The presence of chair conformations have been shown by X-ray crystallographic measurements in substantial number of pyranoid sugars [4].

In this respect radiation damage can be considered from the point of view of its effect rather than its characteristics. Several investigators [5, 6] showed that the radiotic oxidation of vinylpolymers can be studied quantitatively by IR spectroscopy by following the absorption band at 1730 $cm^{-1}$.

*Olaff* et al. [7] studied radiation damage in solid single crystals of organic materials and their structures are discussed in relation to the radiosentizing properties.

For gamma radiolysis *Abou-Sekkina* and *Assar* [8] investigated the radiation damage and hydrogen bonding in some amino-anthraquinone dyes deducing a correlation between the extent of γ-ray damage and the energy of hydrogen bonding. In another more recent publication the present authors [9] studied in details radiation damage and hydrogen bonding in biological active cyclohexane 1,2-diol and pyranol in the solid state. A correlation between γ-radiation damage and the relative strength of hydrogen bonding was put forward and established for the first time [9].

Thus, the major goal of the present manuscript is to evaluate the gamma radiation damage and the interaction mechanism of gamma radiolysis for molecules of monosaccharides (glucose) and disaccharides (lactose) both comprising cyclic structures.

Kurznachrichten / Newlets

Umwelt:
NATUERLICHE RADIOAKTIVITAET

Wussten sie, dass die natuerlichen Radionuklide im menschlichen Koerper bewirken, dass in der Sekunde 37 000 Zerfaelle stattfinden ? Daraus ergibt sich, dass von rund 60 Millionen Menschen ebensoviele radioaktive Zerfaelle stattfinden, wie in 90t Natururan. (Chemie Labor u. Betrieb CLB 36(1985)7)

Kernbrennstoff:
MODELLANLAGE FUER CHEMISCHE U-ANREICHERUNG

In Hyuga-shi in der japanischen Praefektur Miyazaki ist kuerzlich die Modell-U-Anreicherungsanlage der Asahi Chemical Industriy Co. fertiggestellt worden. Die Anlage ist fuer die Anreicherung nach dem chemischen Ionentauscher-Verfahren konzipiert. Ihre Jahreskapazitaet betraegt 1-2t-Trennarbeitseinheiten. Das chemische Anreicherungsverfahren erlaubt die Produktion kleiner Mengen angereicherten Urans. Zudem gilt es als relativ proliferationssicher. Die Anlage in Hyuga-shi, deren Errichtung rund Yen 4 Mia kostete, soll in den kommenden zwei Jahren im Probebetrieb stehen. (SVA-Bulletin (1985)13)

Kernenergie:
SUPERPHENIX KRITISCH

Der Reaktor des Prototyp-Schnellbrueter-Kernkraftwerks Superphenix im franzoes. Creys-Malville oestlich von Lyon wurde am 7. Sept. 1985 erstmals kritisch. Superphenix ist mit einer Leistung von 12oo MW das z.Z. groesste Brueterkraftwerk der Welt und das erste mit einem natriumgekuehlten Schnellen Reaktor ausgeruestete Kernkraftwerk im industriellen Massstab. Betreiberin des Superphenix ist die 1974 gegruendete Gesellschaft Nersa (Centrale nucleaire europeenne a neutrons rapides S.A.), an der franzoesische, italienische, westdeutsche, belgische und niederlaendische Elektrizitaetsunternehmen beteiligt sind. Die Bauarbeiten in Creys-Malville wurden Ende 1976 aufgenommen. Der Nuklearteil der Anlage wurde gemeinsam vom franzoesischen Unternehmen Novatome und der italienischen Firma Ansaldo unter franzoesischer Lizenz in Zusammenarbeit mit mehreren Unterauftragnehmern erstellt. Beim Bau der konventionellen Anlageteile wirkten Unternehmen aus den fuenf Nersa-Laendern sowie der Schweiz mit. Die Aufnahme der Stromerzeugung durch Superphenix ist fuer Anfang 1986 geplatn. (SVA-Bulletin (1985)16)

UEBER 70% ATOMSTROM IN FRANKREICH

In Frankreich kletterte der Atomstromanteil im August 1985 erstmals ueber die 70%-Marke. Gemaess vorlaeufigen Statistiken der Electricité de France (EDF) belief sich die Stromproduktion im Monat August auf 20,58 Mia kWh. Davon stammten 70,7% aus Kernkraftwerken, 20,7% aus Wasserkraftwerken und 8,6% aus konventionell-thermischen Anlagen. Frankreich weist damit den weltweit mit Abstand hoechsten Atomstromanteil auf. Der bisherige Rekordwert von 67,5% war im Dezember 1984 erreicht worden. Im Juli 1985 trugen die franzoesischen Kernkraftwerke 65,0% zur Landesstromproduktion bei. (SVA-Bulletin (1985)16)

WACHSENDE ATOMSTROMANTEILE IM ERSTEN HALBJAHR 1985 IN WESTEUROPA

Nach Mitteilung der Vereinigung Deutscher Elektrizitaetswerke (VDEW) in Frankfurt hat der Anteil der Kernenergie an der gesamten Netto-Elektrizitaetserzeugung im ersten Halbjahr 1984 in verschiedenen europaeischen Laendern weiter zugenommen. So stieg gemaess einer Umfrage der VDEW der Atomstromanteil in Frankreich im genannten Zeitraum gegenueber dem ersten Halbjahr 1984 von 56% auf 62%, in der BRD von 21% auf 30%, in der Schweiz von 33% auf 41%, in Grossbritannien von 15% auf gut 18% und in Italien von 3% auf 4,7%. Leicht ruecklaeufig war der Atomstromanteil in Belgien mit 49% im Vergleich zu 50% im ersten Halbjahr 1984 und in den Niederlanden mit 6,5% gegenueber 7,5%. (SVA-Bulletin (1985) 15)

SCHWERWASSER-ANLAGE IN TALCHER (INDIEN) WIRD BETRIEBSBEREIT ERKLAERT

In dem Bemuehen um eine eigene Schwerwasser-Versorgung hat Indien in den 70er Jahren mit dem Bau dreier kleinerer Schwerwasser-Anlagen mit einer Kapazitaet von je etwa 50 Tonnen pro Jahr begonnen. Neben den Anlagen in Baroda und Tuticorin, die auf franzoesischer Technologie basieren, handelt es sich in Talcher um eine Anlage, bei der das von der Hoechst AG entwickelte Ammoniak-Verfahren angewendet wird. Die Errichtungsarbeiten konnten bereits im Jahre 1982 abgeschlossen werden. Engpaesse in der Stromerzeugung und daraus resultierende ungenuegende Belieferungen mit Ammoniak haben den Probebetriet zur Beseitigung von konstruktiven und operativen Maengeln so sehr verzoegert, dass die Betriebsfaehigkeit der Talcher-Anlagen erst jetzt festgestellt werden konnte. Zwei weitere Schwerwasser-Anlagen in Indien befinden sich in Kota und Thal-Vaishet in Bau. (SVA-Bulletin (1985)15)

Messtechnik:
DURCH KERNPHYSIKALISCHE ANALYSE ERZRESERVEN ERSCHLOSSEN

Da der Bedarf der Volkswirtschaft an mineralischen Rohstoffen staendig steigt und die bekannten Vorraete bereits weitgehend erschoepft sind, muessen Erze mit aermeren Metallgehalt abgebaut werden. Mit Hilfe der kernphysikalischen Analyse mineralischer Rohstoffe entwickelten sie eine Technologie zur Klassierung (nach dem NE-Metallgehalt in den Erzen) und ein System zur Qualitaetssicherung des zu verarbeitenden Haufwerks. Erstmalig in der Welt wurden in einer Reihe von Betrieben der Nichteisenmetallurgie Komplexe automatischer Isotopenanlagen zur Analyse der Erze unmittelbar in den Transportbehaeltern eingefuehrt, wobei letztgenannte nicht gestopt und auch keine Proben entnommen werden muessen. Die Anlagen sind vielfaelltig einsetzbar. Aufgrund der Angaben ueber den Gehalt industriell wertvoller Elemente im Haufwerk koennen taubes Gestein und nicht konditionsgerechte Erze ausgesondert werden. Die Qualitaet des aufzubereitenden Erzes wird besser und der Metallgehalt stabilisiert sich. Dadurch bietet sich die reale Moeglichkeit, arme Erze, darunter von den Halden vergangener Jahre, in den Verhuettungsprozess einzubeziehen. Die Isotopenanlagen sind Bestandteil automatisierter Systeme der Leitung technologischer Prozesse in den Gruben. Wiege- und Messeinrichtungen, Nachrichten- und Fernsehsysteme bilden die exakt funktionierende technische Grundlage eines automatischen Kontroll- und Erfassungssystems fuer die Gewinnung von Erzen und Metallen. Durch die Optimierung des Arbeitsalgorithmus und den Einsatz von Mitteln der Rechentechnik fuer die Bearbeitung der Messergebnisse werden die einzelnen Abteilungen der Bergbaubetriebe termingerecht mit Informationen versorgt. Der Fahrer eines mit Erz beladenen Autokippers erhaelt beispielsweise, wenn er an der Isotopenanlage vorbeifaehrt, auf einem Leuchtbild konkrete Angaben ueber die Erzqualitaet und Hinweise fuer seine weitere Route. Durch den Einsatz des Anlagenkomplexes in den Gruben des Bergbau- und Huettenkombinats Almalyk konnten 1000t Blei und Zink in Konzentraten mit einem oekonomischen Nutzen von mehr als fuenf Millionen Rubel gewonnen werden. Durch die Einbeziehung der Halden einer Lagerstaette des Kombinats in den Verhuettungsprozess auf der Grundlage der Klassierung kann aus der Grube bei gleichbleibender Leistung fuenfzehn Jahre laenger Erz gefoerdert werden. Die neue Technologie erfordert keine grossen Investitionen und ist umweltfreundlich. Sie stellt geringe Anforderungen an die Beschaffenheit der mineralischen Rohstoffe und ermoeglicht die

Einbeziehung armer Erze in den Verhuettungsprozess. (Presse der Sowjetunion (1985)11)

## Veröffentlichungen / Publications

JOINT INSTITUTE FOR NUCLER RESEARCH - DUBNA/USSR

E7-85-431

Half-Lives and Kinetic Energies for the Spontaneously Emitted Heavy Ions from Nuclei

The partial half-lives, branching ratios relative to the alpha decay, kinetic energies and Q-values for the most probable spontaneously emitted heavy ions from 766 nuclides with Z = 47-106 and total half-lives longer than 1 μs are estimated by using the analytical superasymmetric fission model, a semiempirical formula for α-decay life-time, which are not measured, and the Wapstra-Audi new mass tables. Even the socalled "stable" nuclides with Z>40 are meta-stable relative to the new decay modes. The experimental evidences are briefly reviewed. - 1985, D. Poenaru et al.

P13-85-266

On the Registraction of Rare Nuclear Transformations by Detecting Multiple Neutron Emission Events

In the earlier studies of some stony meteorites and geothermal waters there were observed multiple neutron emission events which might be due to the spontaneous fission of superheavy elements (SHE). In this connection it is of interest to look for terrestrial samples with a higher concentration of the sought nuclide, compared with meteorites which represent weakly differentiated solar system matter. A search for such samples can be carried on using an experimental setup weighing 1 ton. Multiple neutron emission detection in rock samples of such a weight will permit the detection of a SHE admixture at a level of $10^{-15}$ g/g after a 20-day exposure. The development and operation of such a setup will allow one to tackle another approach to the experimental problem of nuclear decays due to the nonconservation of baryon number (nucleon decay and $\bar{n}N$ annihilation, as a result of the oscillation of a neutron bound in the atomic nucleons) - 1985, G.N. Flerov et al.

UNITED KINGDOM ATOMIC ENERGY AUTHORITY

AERE R 9271
The AS-76 interlaboratory experiment on the alpha spectrometric determination of 238-Pu - June 1985, 23 pp., E. Foster et al.

AERE-R 10980
Calculations on the feasibility of ultrasonic inspection of steam generator tubing in pressurised water reactors - April 1985, 34 pp., J.A.G. Temple

AERE-R 11182
Studies of environmental radioactivity in Cumbria, part 6. The chronology of discharges of caesium-137, plutonium and americium-241 from BNFL Sellafield, as recorded in lake sediments - April 1985, 42 pp., J.D. Eakins, R.S. Cambray

AERE-R 11574
Specification of a test problem for HYDROCOIN level 1 case 1 : transient flow from a borehole in a fractured permeable medium - April 1985, 32 pp., D.P. Hodgkinson, J. Barker

AERE-R 11404 (Rev.)
A gas dynamic and thermochemical model of steam sodium microleak phenomena - June 1985, 80 pp., R. Perkins, R. Airey, L.C. Daniels

AERE-R 11716
Biological and chemical characterization of landfills (V) Baystone Farm landfill, Sussex - June 1985, 34 pp., S.N. Worrallo

AERE-R 11881
Review of sorption data for site assessment - June 1985, 40 pp., M.H. Bradbury, N.L.Jefferies

SRDP-R125
A doorway monitor for international safeguards - March 1985, 18 pp., R.J. Riley

J A H R E S B E R I C H T E

Philipp-Universitaet Marburg - Jahresbericht 1984/85, 159 S.

Kernforschungsanlage Juelich GmbH - Jahresbericht 1984/85, 160 S.

K A T A L O G E

New Amersham Catalogue - Instrument calibration sources, 1985, 76 pp. - Amersham International, Amersham Place, Little Chalfont, Buckinghamshire, England HP7 9 NA

P R O C E E D I N G S

BHABHA ATOMIC RESEARCH CENTRE, BOMBAY, INDIA

Symposium on Recent Developments in Welding Metallurgy and Processes, February 5 to 7, 1985, held at Bhabha Atomic Research Centre, Technical Papers (total 32 Papers)

CENTRAL INSTITUTE OF ISOTOPES AND RADIATION RESEARCH OF THE ACADEMY OF SCIENCES OF THE GDR, LEIPZIG

Proceedings of the 3rd Working Meeting on "Radiation Instraction" held in Leipzig, GDR, 24 - 28 Sept. 1984, edited by J. Schmidt, O. Brede, R. Doellstaedt and R. Mehnert, 1985, pp. 797 (total 107 papers)

INDUSTRIAL ELECTRON ACCELERATORS

Industrial electron accelerators and their application in chemistry, mechanical industry, biotechnology and medicine are considered in the book. Accelerators with the energy 0,15 - 30 MeV used for applied purposes (d.c. and linear accelerators, betatrons etc.), electrical schemas and designs, principles of action, the main elements of radiation installations (systems for beam extraction, targets, radiation shielding, devices for dosimetry) are described. Chemical processes under the action of ionizing radiation, irradiation of biological objects and also thermal effects of electron beams on matter are discussed. Diagrams and parameters of technological facilities on the basis of electron accelerators (for cross-linking of polyethylene, curing, radiation processing of fabrics, preservation of grain, radial therapy and etc.) are quoted. This book contains parameters' tables of industrial accelerators of different types and about 140 illustrations. The book is intended for specialists designing and servicing the accelerators and also for those, who are developing technological processes with the use of electron beams. It will be useful for students and graduates of appropriate specialities. (Publishing hous: Moscow, Energoatomizdat 1986; Order: 10303, Moscow, Petrovka 15, Technical Book's Shop No. 8)

CHART OF THE NUCLIDES 1984

Japanes Nuclear Data Committee and Nuclear Data Center, Japan Atomic Energy Research Institute, Tokai-Mura, Naga-Gur, Ibaraki-ken, Japan (last edition 1980, esp. actualized at the transuranian elements)

LABORATOIRE DE METROLOGIE DES RAYONNEMENTS IONISANTS

Boite Postale n 21, 91190 Gif-sur-Yvette/France

Problèmes de décroissance radioactive - This monograph is subdivided in three parts. The first one deals with the decav of isolated radionuclides. It includes, in particular, a table of selected half lives, selected values

and corresponding uncertainties and also universal decay tables. The second part deals with the growth of radioactive products. The general relations for two, three and n bodies are given. As an illustration of these relations, some significant cases are thoroughly examined.
The third part deals with activation problems, as well of first order as of the second and in the cases where the primarv radionuclide can be or not be activated. Some relations give induced activity in several different situations. The case of periodic activation is also dealt with - F. Lagoutine, in French

Préparation des sources radioactives -
The measurements of radioactivity and the analysis of radioactive solutions involve the preparation of solid sources. The quality of the sampling is of great importance to the quality of the results. In this monograph, the vanous steps are reviewed:
- methods and accuracies dilutions.
- physico-chemical stability of solutions.
- preparation of different source holders.
- different methods for aliquote determination (volumenometry, microweighing) and associated uncertainties.

Phenomena inherent to the measurement of such sources (absorption, retrodiffusion, autoabsorption) and corrections to be considered are studied too - Yves Le Gallic, in French

TABLE OF RADIONUCLIDES

Ten years ago, the Laboratoire de Métrologie des Rayonnements Ionisants (LMRI) decided to publish a table of radionuclides, the data of which would be evaluated according to well defined rules. Since, two volumes of this table and a "sélection medicale" have been published. The success encountered by this document and the need for updating led the LMRI to publish a new issue and to carry on evaluation work (3rd volume):
1st volume, edition 1983; 2nd volume, edition 1984; 3rd volume, edition 1983-1985
In this issue, as in the previous one, an uncertainty is associated with each parameter (half-life, energy, and intensity of each radiation) and is defined for a given confidence level. Furthermore, in a pragmatic effort, the LMRI has added to the decay scheme the main emission spectra and decay tables of the radionuclide in question. A summary of bibliographic references is provided so as to enable the reader to refer to source materials.
The 1983 edition of the "sélection medicale" provides in addition, values of the mean energy emitted per nuclear transformation - the equilibrium dose constant, $\Delta$ - for the main emissions of radionuclides considered.
Each radionuclide is considered in a single booklet wich comprises the following:
- decay schema
- tables of nuclear, atomic and radiation data
- major production methods
- references employed in evaluation
- decay table
- $\gamma$ X-ray and conversion electron spectra

Furthermore, in the Ist volume, the tables are preceded by a short description of the radioactive processes and of the evaluation rules used. In the other volumes, summaries of the symbols and definitions are included.

## Veranstaltungskalender — Tagungen — Lehrgänge / Calendar — Meetings — Courses

WISSENSCHAFTLICHE TAGE DER TECHNISCHEN HOCHSCHULE "OTTO VON GUERICKE" Magdeburg
Im Rahmen der Wissenschaftlichen Tage 1986 der Technischen Hochschule "Otto von Guericke" sind u.a. auch folgende Veranstaltungen vorgesehen:
- 11. Projektierungskolloquium der Betriebs- und Arbeitsgestaltung - 2./3.9.1986

Veranstalter: Sektion Technologie der metallverarbeitenden Industrie
Thema: Arbeits- und betriebsgestalterische Aspekte der flexiblen Automatisierung im Maschinenbaubetrieb

14. Tribotechnisches Kolloquium Waelzlagertechnik - 9./10.9.1986
Veranstalter: Sektion Maschinenbau Wissenschaftsbereich Maschinenbautechnik
VEB Kombinat Waelzlager und Normteile, BT Forschung u.Entwicklung
Thema: Hochbeanspruchte Waelzpaarung und Waelzlager
Themenkomplexe:
- Grundlagen fuer Waelzpaarungen
- Waelzlager
- Schmierung von Waelzpaarungen und Waelzlagern

4. Fachtagung "Anwendung von Mikrorechnern in der Mess- und Automatisierungstechnik" - 11./12. 9.1986
Veranstalter: Sektion Technische Kybernetik und Elektrotechnik, Forschungsgemeinschaft "Anwendung der Mikroelektronik im Schwermaschinen- u. Anlagenbau
Schwerpunkte:
- Messwerterfassungsstrukturen fuer die Prozessmesstechnik
- intelligente Funktionseinheiten zur Aggregatautomatisierung
- prozessnahe Interfaceelektronik
- Mikrorechnergeraetetechnik zur Messwerterfassung, -uebertragung und vorverarbeitung einschl. schaltungstechnischer Probleme
- Fragen zur Analogwerteingabe und Messwertvorverarbeitung
- Sensor-Anwendungen und Sensoreingangsschaltungen
- Entwicklung und Anwendung moderner Algorithmen der automatischen Steuerung und ihre Umsetzung in Automatisierungsloesungen mit Mikrorechnern in Echtzeitverarbeitung
- Probleme der Geraete- und Anlagentechnik, ihre Dimensionierung und Projektierung
- Darstellung ausgewaehlter Anwendungsfaelle der Regelungstechnik und Prozess-Steuerung vornehmlich aus dem Bereich des Maschinen- und Anlagenbaues sowie der Verfahrenstechnik.

Naehere Auskuenfte erteilt: Technische Hochschule "Otto von Guericke Magdeburg - Tagungsbuero - DDR-3010 Magdeburg, PSF 124

XXXVII. Berg- und Huettenmaennischer Tag
Bergakademie Freiberg - 24.-27.6.1986

Rahmenthema: "Suche, Erkundung, Gewinnung und Verarbeitung von mineralischen Rohstoffen und Energietraegern sowie Fragen der Werkstofftechnik und der Energietechnik".

Im Rahmen des Berg- und Huettenmaennichen Tages 1986 ist auch ein Kolloquium ueber "100 Jahre Germanium"; im Februar 1986 jaehrt sich zum 100. Mal die Entdeckung des Elements Ge durch den Freiburger Chemiker und Professor an der Bergakademie, Clemens Winkler.

6. Wissenschaftliche Konferenz ueber "Zerstoerungsfreie Pruefung im Bauwesen", 2.-4.12.1986 in Leipzig

Veranstalter: Technische Hochschule Leipzig, Sektion Naturwissenschaften, DDR-7030 Leipzig, PSF 66

## Ausgewählte Beiträge / Recently published papers

ATOMNAJA ENERGIJA 60(1986)1 (all in Russian)
Problems of creating regional systems for radiation situation monitoring in areas affected by nuclear facilities - I.S. Eremeyev

Restoration of neutron spectra from activation measurements using the transport equation - V.D. Lartsev et al

The role of Tc in oxidation-reduction technological processes of irradiated fuel - V.S. Koltunov et al

Some applications of the resonant light ion backscattering spectrometry method - V.S. Shorin, A.N. Sosnin

Cross sections of fast-neutron forward scattering by U-238 and Pu-239 nuclei - G.V. Anikin, I.I. Kotukhov

Cross sections of fast-neutron forward scattering by Cu and Pb nuclei - G.V. Anikin, I.I. Kotukhov

Implantation kinetics of accelerated gas ions in crystallic solids - A.B. Kitsenko, N.P. Katrich

On the possibility of X-ray fluorescense determination of Tc in solutions - G.A. Akopov et al

Absolute measurements of the prompt neutron spectrum for spontaneous fission of Cf-252 in the 0,020-1,220 MeV energy range - P.P. Dyachenko et al

Neutron yield stabilization for neutron generator NG-150 - L.E. Kuzmin et al

Fission cross-section of long-lived isomer Np-236 for thermal neutrons - B.N. Belyayev et al

DOKLADY AKADEMII NAUK 281(1985)5

On the calculation of the efficiency of $\alpha$-particle irradiation use for the pre-ionization of working medium in gas-discharge lasers and amplifiers - N.G. Basov et al.

On the quantitative estimation of the Earth degassing dynamics - V.N. Kortsenstein

DOKLADY AKADEMII NAUK 282(1985)1

Positron annihilation at polycrystal grain bounderies - I.J. Dekhtior et al.

Fractionation of C isotopes by thermophilic methane-forming bacteria - M.V. Ivanov et al.

PLANT AND SOIL 83(1985)2

Distribution and recovery of N-15 after fertilization of Douglas-fir saplings with different N sources - P.C.K. Pang

Effect of mild drying on the mineralization of soil N - R. Seneviratne et al.

PLANT AND SOIL 83(1985)3

Fate of N applied as Azolla and blue-green algae (Cyanobacteria) in water logged rice soils - a N-15 tracer study - M.H. Mian et al.

A N-15 tracer study to compare N supply by Azolla and $(NH_4)_2SO_4$ to IR 8 rice plants grown under flooded conditions - M.H. Mian et al.

THE ASTROPHYSIC. J. 290(1985)1 Part 1

The Li isotope ratio in 5F-dwarfs or G-dwarfs - L.M. Hobbs

SEPARATION SCI. TECHNOL.19(1984)8/9

Theory of isotope separation by electrochemical parametric pumping - Y. Oren et al.

Z. PHYSIK. CHEM. (Leipzig) 266(1985)1

Photodissociation of $CH_2O$ - calculation of isotopic effects in the dissociation probabilities - K. Jost et al.

CANAD. J. EARTH SCI. 22(1985)1

Sm/Nd isotopic investigation of the age and origin of the Meguma Zone metasedimentary rocks - D.B. Clarke et al.

Radiocarbon age of rock-boring Hiatella arctica (Linné) and postglacial sea-level change at Cow head, New Foundland - J.A. Brookes et al.

ATMOSPHERIC ENVIRONMENT 19(1985)1

Rn transport into a detached one-story house with a basement - W.W. Nazaroff

Long-range atmospheric dispersion of radioisotopes Part I and II - H.M. Ap Simon et al.

DEEP-SEA RES. 32(1985)1

The distribution of C-13 of $\Sigma CO_2$ in the world oceans - P.M. Kroopnick

A comparison of N assimilation rates based on N-15 uptake and autotrophic protein synthesis - E.A. Laws et al.

ADVANCES IN GEOPHYSICS 26(1984)

Climate model simulation of $CO_2$-induced climatic change - M.E. Schlesinger

NATURE 314(1985)6009

Pb isotope differences between whole-rock and phenocryst in recent lavas from southern Italy - M. Cortini et al.

NATURE 314(1985)6010

Stable isotope data and depositional environments in the late Quaternary Arctic Ocean - R. Zahn et al.

J. AMER. CHEM. SOC. 107(1985)6

Kinetic isotope effects for H abstraction from a series of cycloalkanes and branched alkanes by H atoms in the gaseous phase - N. Fujisaki et al.

THE INTERNAT. J. APPL. RADIATION ISOTOPES 36(1985)2

Radiological impact of fallout Tc-99 in Seville, 1965-1967 - M. Garcialeon et al.

Substitution reactions of Tc N $Cl_4$-Tc-99m -a route to a new class of Tc-99m - radiopharmaceuticals - J. Baldas et al.

Liquid-scintillation counting efficiency as a function of the figure of merit for pure $\beta$-particle emitters - A.G. Malonda et al.

The $\beta$-ray gauge as an evaporimeter - N.N. Barthakur

Ir-192 concentration on the eluate of the Os-191-Ir-191m generator - A.B. Packard et al.

The use of $CO_2$ absorbers for the determination of specific C-14-activities - R.M. Qureshi et al.

GEOKHIMIYA (1985)2 (in Russian)

K-U-Th systematics of the matter of solar system planetary bodies - A.T. Basilevsky

Distribution of stable and radioactive isotopes of K in waters of the Western Pacific - V.V. Anikeev et al.

New data of S isotopy in hydrothermal waters of the Mendeleev Volcano-Kunasiv Island - J.P. Averyanov et al.

Mass-spectrometrie determination of output of Xe isotopes in reactions of deep splitting of Er, Tu, Ta, W and Re by protons with 1GeV energy - D.V. Mink et al.

GEOCHIM. COSMOCHIM. ACTA 49(1985)3

Isotopic composition of noble gases in geothermal fluids of the Krusne-Hory Mts. Czechoslovakia, and the nature of the local geothermal anomaly - B.G. Polyak et al.

Distribution and isotopic composition of S in lake sediments of Northern Ontario - J.O. Nriagu

## Experimental

a) *Materials.* i) Glucose; $C_6H_{12}O_6$, pure analytical grade supplied from El-Nasr Pharmaceutical Chemical Co., Abu-Zaabal, Egypt. M.W. 180·16 anhydrous. ii) Lactose; $C_{12}H_{22}O_{11} \cdot H_2O$. Pure analytical grade supplied from El-Nasr Pharmaceutical Co., Abu-Zaabal, Egypt.

b) *Infrared Absorption Measurements.* The IR spectra of the materials investigated were carried out at room temperature using the solid KBr disc technique. A Perkin-Elmer (W. Germany) 683 infrared spectrophotometer was employed in the range of 4000 – 200 $cm^{-1}$.

c) *Electronic Absorption Spectra Measurements.* The nujol mull solid technique was followed for obtaining the room temperature electronic absorption spectra of the test samples. A Unicam SP 8000 ultraviolet recording spectrophotometer was utilized in the range 200 – 450 nm.

d) *γ-Radiation Exposure.* The materials were exposed to an absorbed gamma dose of $5{\cdot}0 \times 10^7$ rd in the "Middle Eastern Regional Radioisotope Centre for the Arab Countries, Dokki, Cairo, Egypt" using $^{60}$Co-gamma cell (activity 300 Ci) in air at room temperature.

## Results and discussion

Fig. 1, curve B shows the room temperature infrared absorption spectrum of glucose molecule before gamma irradiation and after $5{\cdot}0 \times 10^7$ rd (curve A). After irradiation the spectra display the presence of saturated aliphatic groups, hydroxyl groups, difinite new band at 2325 $cm^{-1}$, commonly associated with aldehydic groups [10], another new band appeared at 1700 $cm^{-1}$ seems to originate from a carbonyl group (see Fig. 1, curve A). A moderately strong but definite band at 1690 $cm^{-1}$ commonly associated with aliphatic double bonds, all of these can be explained on the basis of the keto-enol tautomerism. It was also found that the intensity of the band at 1100 $cm^{-1}$ assigned to secondary alcohols is decreased.

These observations appear to establish the formation of the carbonyl groups after gamma irradiation. These together with the appearance of colour change of the test sample from colourless to yellowish by irradiation indicates the gamma radiotic oxidation cleavage (opening of the pyranose ring of glucose [11]) and the cyclic structure of glucose molecule was destroyed.

For lactose molecules, which contain one molecule of glucose and another molecule of galactose, the infrared

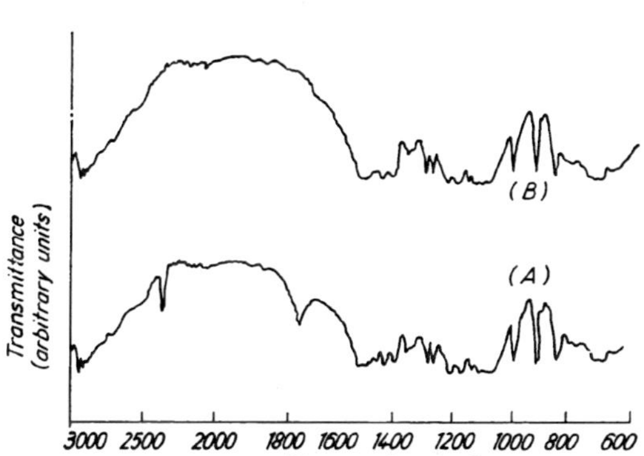


Fig. 1. The room temperature infrared absorption spectra of glucose molecules
B) before irradiation; A) after gamma irradiation ($5{\cdot}0 \times 10^7$ rd)

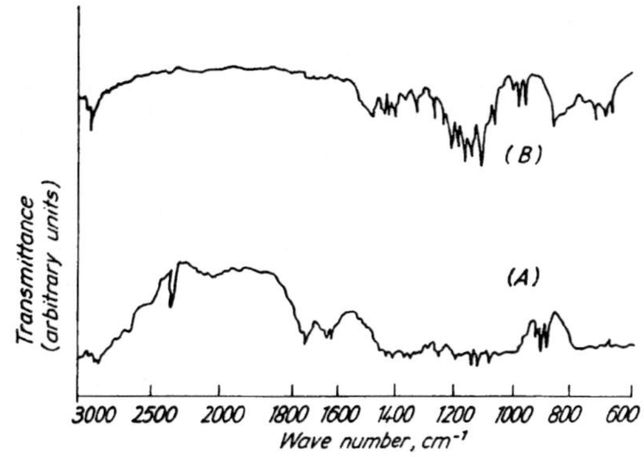


Fig. 2. The room temperature infrared absorption spectra of lactose molecules
B) before irradiation; A) after gamma irradiation ($5{\cdot}0 \times 10^7$ rd)

Tab. 1. Band assignments of the characteristic absorption bands of glucose and lactose molecules before (B) and after (A) the absorbed gamma dose ($5{\cdot}0 \times 10^7$ rd)

| Compound | M.P. | IR absorption bands $cm^{-1}$ | | Band assignment | Ref. |
|---|---|---|---|---|---|
| | | B | A | | |
| | | | 2325 | Associated with aldehydic group | [10] |
| Glucose | 90 | 2040 | 2040 | | |
| | | | 1700 | $\nu_{C=O}$ | |
| | | | 1690 | $\nu_{C=C}$ | |
| | | | 1100 smaller | 2° = OH | |
| Lactose | 210 | | 2325 | Associated with aldehydic group | [10] |
| | | | 2050 | | |
| | | | 1730 | $\nu_{C=O}$ | |
| | | | 1690 | | |
| | | | 1650 | | |
| | | 1100 | 1100 smaller | 2° = OH | |

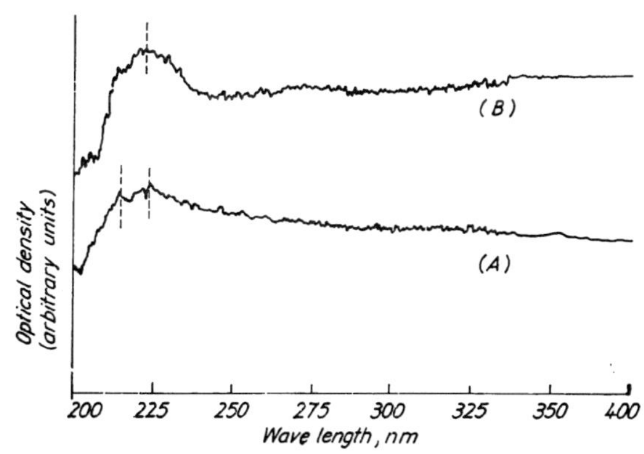


Fig. 3. The room temperature electronic absorption spectra of glucose molecules
B) before irradiation; A) after gamma irradiation ($5{\cdot}0 \times 10^7$ rd)

absorption spectrum after the absorbed gamma dose (Fig. 2, curve A) shows the presence of saturated aliphatic groups, hydroxyl groups and a strong definite band at 2325 $cm^{-1}$ commonly associated with aldehydic group.

Another new bands at 1730 $cm^{-1}$ and 1650 $cm^{-1}$ were appeared after irradiation, which seems to originate from a carbonyl groups (C=O) for glucose and galactose molecules.

This explain that the cyclic structures of glucose and galactose in lactose molecule was opened and evidence for these, is the existence of colour change of the test sample from colourless to yellowish brown colouration as a result of energetic gamma absorbed dose ($5{\cdot}0 \times 10^7$ rd) (see Tab. 1).

This ensured that the cyclic pyranose structures of glucose and galactose molecules were opened as a result of exposure to energetic gamma absorbed dose ($5{\cdot}0 \times 10^7$ rd) or gamma radiation damage.

The ultraviolet absorption spectrum of glucose is shown in Fig. 3 before (B) and after (A) the absorbed gamma dose ($5{\cdot}0 \times 10^7$ rd). Assignments of the most characteristic bands were also listed in Tab. 2. After irradiation (curve A) the strong absorption band which has $\lambda_{max}$ at 220 nm, is splitted into two bands at 215 nm and 225 nm which commonly associated with aldehydic carbonyl group [12].

This could be explained on the basis of the existence of Keto-enol tautomerism and could plausibly be correlated with the opening of the cyclic structure of glucose after exposure to γ-radiation. This mechanism may take place through gamma radiotic oxidation of the hemiacetal structure to aldehydic structure (see formula I).

Evidence supports this idea is taken from band assignments at 215 nm (see Fig. 3 and Tab. 2).

Fig. 4 shows the ultraviolet absorption spectrum of lactose (B) before gamma irradiation and (A) after the absorbed gamma dose ($5{\cdot}0 \times 10^7$ rd).

It is of interest to note that after irradiation the band at 220 nm splitts into 215 and 225 nm as a result of gamma radiation exposure. This confirms that previously concluded in the case of glucose molecule as being due to the opening of the two lactose cyclic rings (see formula II) through the radiotic oxidation of the hemiacetal structure to the aldehydic structure displaying two aldehydic absorption bands at 215 and 225 nm (see Fig. 4 and Tab. 2).

Tab. 2. Band assignments of the characteristic electronic absorption bands of glucose and lactose molecules before (B) and after (A) the absorbed gamma dose ($5{\cdot}0 \times 10^7$ rd

| Compound | Electronic absorption in nm | | Band assignments | Ref. |
|---|---|---|---|---|
| | B | A | | |
| Glucose | 220 | 215<br>225 | Electronic transitions n − π | [10, 12] |
| Lactose | 220 | 215<br>225 | Electronic transitions | [10, 12] |

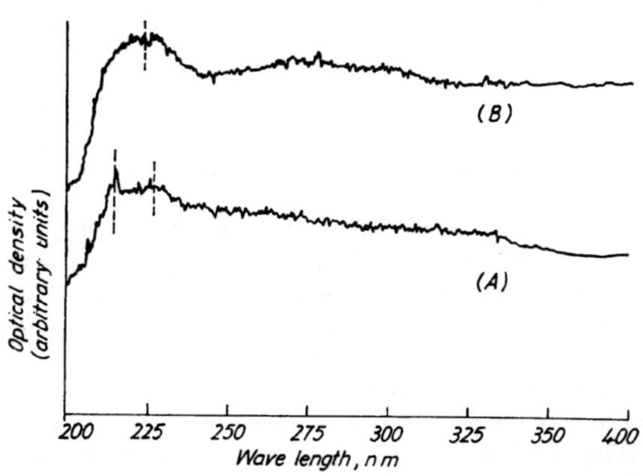


Fig. 4. The room temperature electric absorption spectra of lactose molecules
B) before irradiation; A) after gamma irradiation ($5{\cdot}0 \times 10^7$ rd)

## Conclusions

Accordingly, from discussions attained at the chemical changes undergone by irradiation of glucose and lactose emphasis that gamma radiotic oxidation reaction accounts foreby the production of some carbonyl functions such as in diketone in 2, 3 positions in the glucose and lactose molecules accompaning the irradiation interaction in both molecules. The nature of these carbonyl groups have not been extensively examined, but the evidence available at present indicates that both diketone groups in 2,3 positions are formed. The diketone functional groups were proved from the reversible colour changes (yellow to colourless) that the irradiated glucose and similarly lactose will undergo on alternate drying and exposure to moisture. The yellow colour of the irradiated glucose and lactose samples is present in a conjugated system such as in benzil [13] molecule which is yellow in colour. The colourless forms of the irradiated glucose and lactose molecules took place by hydration of the ketone form which would destoy the conjugation as follows:

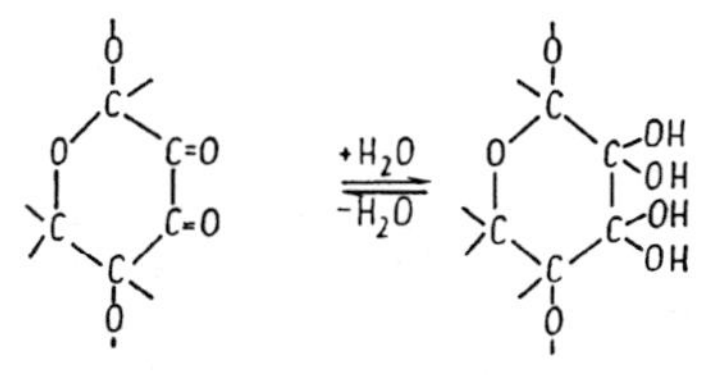


yellow colourless

Alternatively our results of IR and electronic absorption spectra obtained were in full agreement and hold good exactly with one another.

This supports the following suggested mechanism of gamma radiation interaction with either glucose (as an e.g. of monosaccharide) or lactose (as an e.g. of disaccharide) molecules: I:

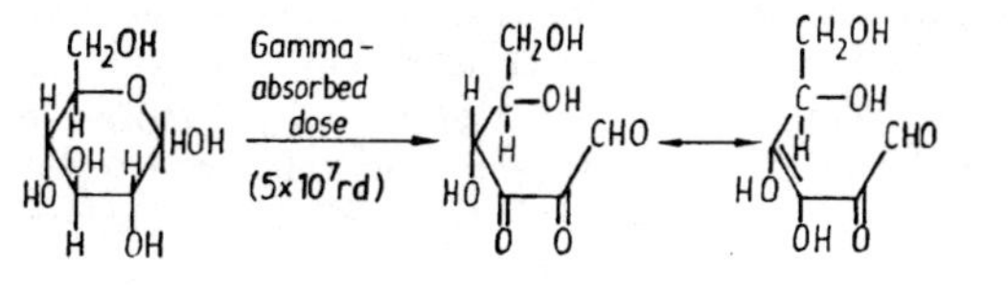


II:

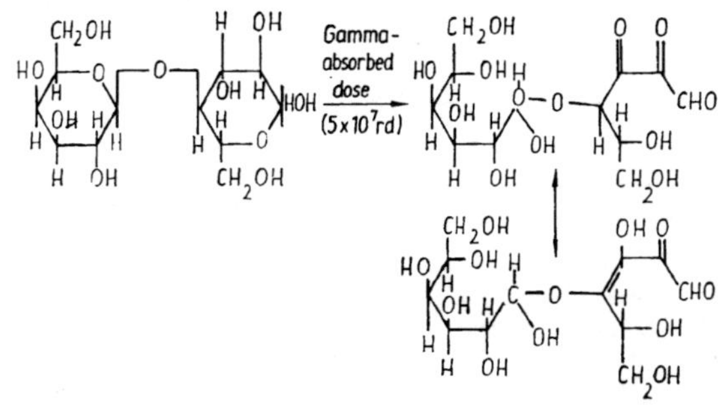


It is of interest to note that for the latter case (in lactose) the gamma radiotic oxidation may also take place either on glucose ring or on galactose or on both.

Received January 22, 1985
Accepted in revised form May 28, 1985

## References

[1] *O. Hassel, B. Ottar*, Acta Chem. Scand **1** (1947) 929

[2] *R. E. Reeves*, (a) JACS **71** (1949) 215; (b) JACS **72** (1950) 1499; (c) Advan. Carbohydrate Chem. **6** (1951) 107

[3] *H. S. Isbett*, J. Res. Natl. Bur. Std. **57** (1956) 171

[4] a) Sucrose: *C. A. Beevers, T. R. R. McDonald, F. Robertson*, Acta Cryst. **5** (1952) 689
b) α-D-glucosamine hydrobromide- *E. G. Cox, G. A. Jeffrey*, Nature **143** (1939) 894
c) α-D-glucose: *T. R. R. McDonalds, C. A. Beevers*, Acta Cryst. **5** (1952) 654
d) B)-Cellobiose: *R. A. Jacobson, J. A. Wunderlich, W. N. Lipscomb*, Acta Cryst. **14** (1961) 598

[5] *D. S. Ballantine, G. J. Diens, Morwitz*, BNL (1953) 1677

[6] *R. M. Blach, A. Charlesby*, Int. J. Appl. Rad. Isotopes **7** (1959) 134

[7] *H. Olaff, E. Haindi, J. Huttermann, J. Kraus*, Radiation Research **80** (1979) 447

[8] *M. M. Abou-Sekkina, S. S. Assar*, Int. J. Appl. Rad. **32** (1981) 847

[9] *M. M. Abou-Sekkina, M. A. Sakran, A. A. Saafan*, Isotopenpraxis **21** (1985) 281

[10] *B. J. Hendrickson, J. D. Cram, S. G. Hammond*, "Organic Chem." 3rd Edition, McGraw-Hill 1970, 757

[11] *M. A. Moharram, S. M. Rabie*, Isotope Rad. Res. **10** (1978) 91

[12] *H. Becker, W. Berger, G. Domscke*, Organikum. Berlin: VEB Deutscher Verlag der Wissenschaften 1974, 98

[13] *T. P. Nevell*, J. Textile Inst. **39** (1948) 118


Isotopenpraxis 22 (1986) 3, S. 93 bis 96

# Effects of Temperature and Gamma Radiation Damage on the Stress-Strain Characteristics of Some Al-Fe and Al-Si Alloys

*M. M. Abou Sekkina*[1])

(Chemistry Department, Faculty of Science, Tanta University, Tanta, Egypt)

*M. R. Nagy, M. Sakr*

(Physics Department, Faculty of Education, Ain Shams University, Egypt)

*The effects of gamma irradiation have been extensively investigated by stress-strain measurements in the temperature range from 300 K up to 475 K for Al—Fe and Al—Si alloys. Doping of small concentrations of 0·0005—0·0030 at. % Fe and 0·01—0·37 at. % Si, in Al-matrix caused a decrease in the work hardening coefficient K, yield stress $\sigma_y$ and fracture stress $\sigma_f$, while the stress-strain parameters observed were increased after the gamma irradiation process. The decrease of stress-strain parameters with increasing both temperature and proportion of alloying elements were attributed to the migration of vacancy-atom piars in the Al-matrix. After gamma irradiation the increase of the stress-strain parameters was highly attributed to atomic displacements and the interaction of dislocations formed by irradiation process with Si- and Fe-vacancy pairs.*

*Es wurden Gammastrahleneffekte durch Messung von Zugdehnungseigenschaften für Al—Fe- und Al—Si-Legierungen im Temperaturbereich von 300 ··· 475 K untersucht. Die geringen Zusätze von Fe (0,0005 ... 0,0030 at-%) und Si (0,001 ... 0,37 at-%) verursachten in der Al-Matrix eine Abnahme des Kältehärtungskoeffizienten K, der Fließspannung $\sigma_y$ und der Bruchspannung $\sigma_f$, während nach einer Gammabestrahlung die Zugdehnungsparameter eine Zunahme zeigten. Die Abnahme der Zugdehnungsparameter bei zunehmender Temperatur und höheren Zusätzen wurde der Migration von Leerstellen-Atompaaren in der Al-Matrix zugeschrieben. Eine Zunahme der Zugspannungsparameter nach einer Gammabestrahlung wurde auf die Verlagerung von Atomen und auf Wechselwirkungen der Versetzungen, die sich bei der Bestrahlung bilden, zurückgeführt.*

## Keywords

aluminium alloys; atomic displacements; gamma radiation; nuclear damage; radiation effects; strain aging; stresses; temperature dependence

## Introduction

Alloys of aluminium and silicon form an important part of aluminium scrap. Liquid quendhing [1] as compared to solid quenching enlarges the solid solubility of the alloying elements, thus facilitating the recycling of scrap material. The precipitation of silicon in solid quenched Al—Si alloys was the subject of several studies [2, 3].

Iron and to some extent silicon are impurities of duralumin alloys. Iron is a harmful impurity. By forming the compound (Mn, Fe) $Al_6$, iron considerably reduces the strength and ductility of duralumin. Therefore, iron content should never exceed 0·50—0·60 percent [4].

[1]) Author for correspondence to be addressed

The γ-irradiation of Al-alloys rises the density of lattice defects in Al-matrix and accordingly decreases their mobility. Stress-strain characeristics of some Al-alloys have been previously studied and implied that the strength and ductility of Al-alloys were due to the interaction between the solute atoms and the lattice defects existing during stress-strain test. Several investigations were carried out to interpret the effect of γ-irradiation on the mechanical characteristics of several metals and alloys. The γ-irradiation effect gives rise to the strengthening and work hardening of the irradiated alloys.

The effect of deformation temperature on the stress-strain characteristics of the irradiated Al-alloys, is to return back in steps the stress-strain parameters to their initial values. This effect seems to be reasonable for the disappearance of the lattice defects as created by γ-irradiation, due to the mutual recombination and annihilation of the irradiation defects at their sinks at temperatures where the defects become appreciably mobile.

The present work aims to determine the effect of γ-irradiation and reverse effect of temperature on the stress-strain characteristics of both Al–Fe and Al–Si alloys which have not yet tackled before.

## Experimental

The investigated compositions of Al–Fe and Al–Si (BDH, England) are spec. pure Al (99·9999 at.%), Al – 0·0005 at.% Fe, Al – 0·0015 at.% Fe, Al – 0·0030 at.% Fe; Al – 0·01 at.% Si; Al – 0·02 at.% Si; Al – 0·11 at.% Si; and Al – 0·37 at.% Si, respectively. The samples are wires of diameters $4 \times 10^{-4}$ m. They were annealed at 773 K for $10^4$ s to eliminate the defects introduced during swaging. The alloys were irradiated using $^{60}$Co γ-cell (activity 300 Ci) at room-temperature. The samples were exposed for an absorbed dose of $0{\cdot}58 \times 10^7$ rd. at the Arab Institute of Radioactive Isotopes, Cairo, Egypt.

Before and after γ-irradiation stress-strain tests were performed on samples of gauge length $5 \times 10^{-2}$ m with a constant strain rate $\dot{\varepsilon} = 7 \times 10^{-6}$ s at temperature range from 300 K to 474 K.

## Results

Typical stress-strain curves of annealed and γ-irradiated Al–Fe and Al–Si alloys are illustrated in Fig. 1 and 2, respectively. It was noticed that the strength of the samples is decreased with increasing both the working temperature and the alloying elements. After γ-irradiation, the strength of the test samples was raised and compensated with rising

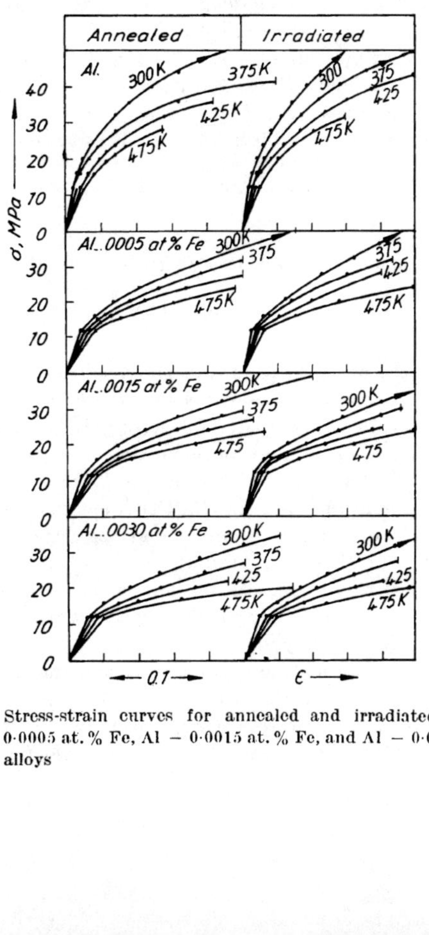


Fig. 1. Stress-strain curves for annealed and irradiated Al, Al – 0·0005 at.% Fe, Al – 0·0015 at.% Fe, and Al – 0·0030 at.% Fe alloys

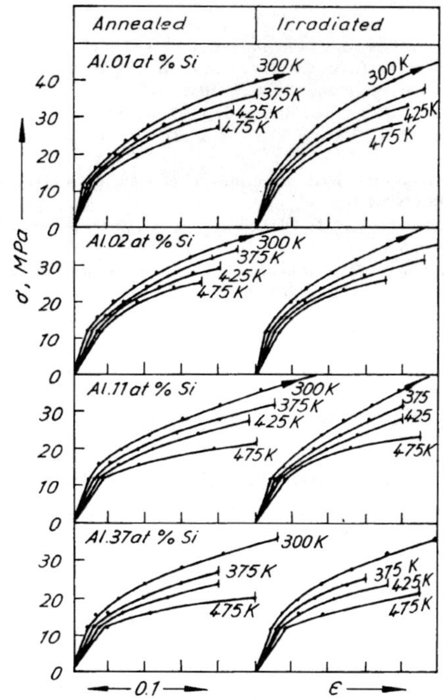


Fig. 2. Stress-strain curves for annealed and irradiated Al – 0·01 at.% Si, Al – 0·02 at.% Si, Al – 0·11 at.% Si and Al – 0·37 at.% Si alloys

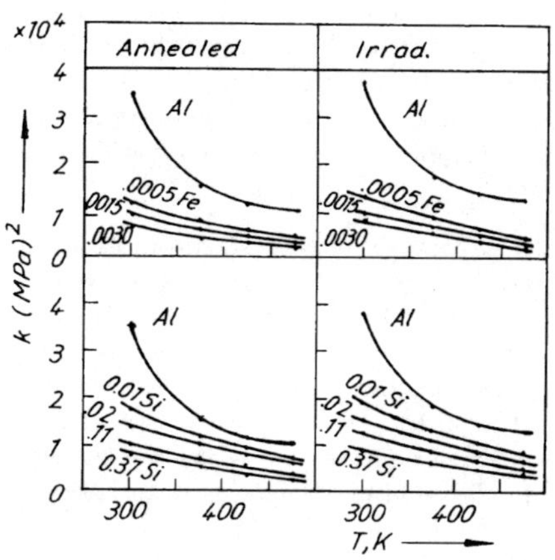


Fig. 3. Effect of working temperature on the coefficient of work hardening (*K*) of annealed and irradiated Al, Al – 0·0005 at.% Fe, Al – 0·0015 at.% Fe, Al – 0·0030 at.% Fe, Al – 0·01 at.% Si, Al – 0·02 at.% Si, Al – 0·11 at.% Si and Al – 0·37 at.% Si alloys

the deformation temperature. These curves show a parabolic stress-strain relations in the intermediate range of stress and strain. The parabolic work hardening coefficient $\left(K = \frac{\partial \sigma^2}{\partial \varepsilon}\right)$ was determined using *Mott*'s model [5]. It was found to be characterised to the working temperature, the solute atoms and γ-irradiation. From Fig. 3, it can be seen that *K* decreases with increasing both the deformation temperature and the concentration of the alloying elements. After γ-irradiation values of *K* were increased.

The observed values of both yield and fracture stresses ($\sigma_f$ and $\sigma_y$) for both alloys before and after γ-irradiation were found to decrease monotonically with rising both the working temperature and the percentage of the alloying

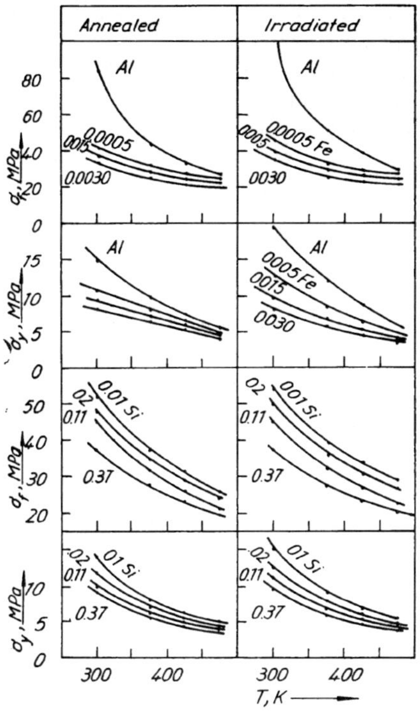


Fig. 4. Effect of working temperature on the fracture stress $\sigma_f$ and yield stress $\sigma_y$ annealed and irradiated Al, Al – 0·0005 at.% Fe, Al – 0·0015 at.% Fe, Al – 0·0030 at.% Fe, Al – 0·01 at.% Si, Al – 0·02 at.% Si, Al – 0·11 at.% Si and Al – 0·37 at.% Si alloys

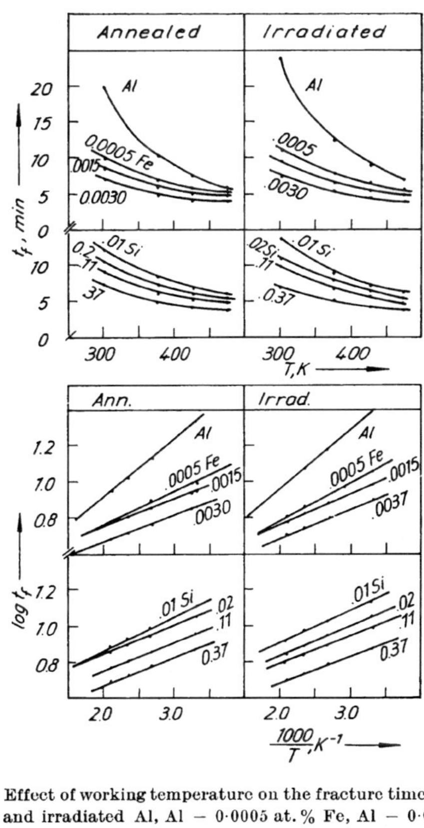


Fig. 5. Effect of working temperature on the fracture time $t_f$ of annealed and irradiated Al, Al – 0·0005 at.% Fe, Al – 0·0015 at.% Fe, Al – 0·0030 at.% Fe, Al – 0·01 at.% Si, Al – 0·02 at.% Si, Al – 0·11 at.% Si and Al – 0·37 at.% Si alloys

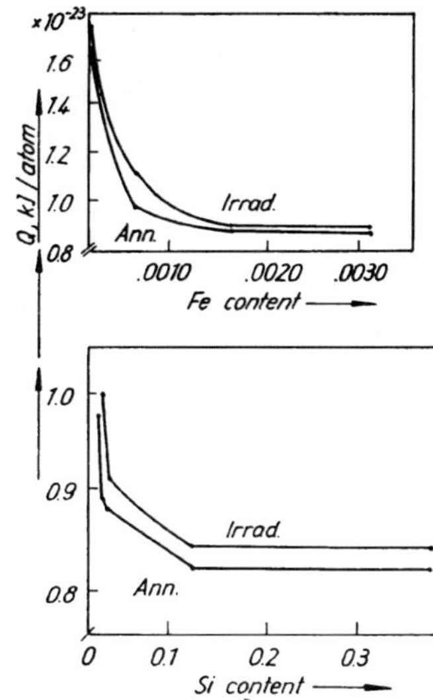


Fig. 6. Variation of the activation energy $Q$ with iron content and silicon content for Al – Fe and Al – Si alloy respectively

elements. Obviously, the effect of γ-irradiation resulted in an increase of both $\sigma_y$ and $\sigma_f$ as shown in Fig. 4.

The temperature dependence of the fracture time ($t_f$) for both alloys before and after γ-irradiation is shown in Fig. 5. It is clear that $t_f$ decreased with increasing the percentage of the alloying elements. It was also found that it is increased after γ-irradiation.

The activation energy of fracture mechanism for both alloy systems was determined from the slope of the straight lines of ln $t_f$ and $1000/T$ (see Fig. 6). It was found to be decreased with increasing minute percentages of Fe and Si, and increased after γ-irradiation.

## Discussion

The decrease in the coefficient of work hardening $K$, the yield stress $\sigma_y$ and the fracture stress $\sigma_f$ with increasing minute percentages of the alloying elements can be explained as a result of the interaction of Fe and Si substitutional atoms with Al vacancies to form atom-vacancy pairs (Fe-vacancy pairs and Si-vacancy pairs). These pairs tend to enhance the motion of the mobile dislocation to continue glide during stress-strain tests. Such continuity produces on ordered flow stress fields that facilitated further glide of dislocation and leads to decreasing in the strength of the Al-matrix.

The increase in the stress-strain parameters ($K$, $\sigma_y$ and $\sigma_f$) after γ-irradiation is thought to be due to the creation of new lattice defects acting as new obstacles which pin the dislocations and rotard their migration in the irradiated Al-matrix. This leads to more strengthening for both alloys.

The deformation temperature exhibits a reverse effect of γ-irradiation on the stress strain parameters. It agitates thermally the dislocation motion and lowers the strength of both alloys. The observed decrease in the activation energy with increasing the minute concentrations of the alloy elements might be as a result of agitation of dislocation glide by the Fe and Si-vacancy pairs. After γ-irradiation, the observed increase in the activation energy might be as a result of increasing the density of new formed stable defects which retard and immobilize the dislocation glide during the stress-strain tests. The fracture mechanism in both alloy systems was found to be dislocation intersection as it has an activation energy ranged from (0·85 to 1·8) × $10^{-22}$ kJ/atom. These values are consistent with the previous values [6]. It was found also to depend on the minute concentrations of alloying elements and γ-irradiation.

Alternately, the primary cause of radiation damage is the displacement of atoms and electrons. In metals, only the displacement of atoms is thought to be important. In some cases, the displacements of electrons may lead to the displacement of atoms. Displacement of atoms can occur directly from the transfer of sufficient momentum from the bombarding particle. Ideas on the nature of clusters of vacancies have developed rapidly in recent years. *Bartlett* and *Dienes* [7] suggested that divacancies would be more mobile than single vacancies. In accordance with *Damask* et al. [8] on workin on copper, the present gamma radiation damage induced large relaxation with strong binding and very low mobility in dil Al—Si and Al—Fe alloys. All of these may result in the observed increase of stress-strain parameters in the investigated Al-alloys.

Accordingly, gamma irradiation finds its useful application in modifying the stress-strain parameters of Al-alloys, on using very small proportion of either Si or Fe dopants. All of these were inexpensive for the great Al-industries.

Received May 15, 1985

### References

[1] *A. Bendijk, R. Delhez, L. Katgerman, Th. H. van de Keiser, E. J. Mittemeizer, N. M. Jan der Pers*, J. Mater Sci. **15** (1980) 2803
[2] *E. Ozama, H. Kimura*, Acta Metall. **18** (1970) 995
[3] *E. Ozama, H. Kimura*, Mater. Sci. Eng. 8 (1971) 327
[4] *Y. Lakhtin*, Engineering Physical Metallurgy. Moscow: Mir Publisher 1971, 400
[5] *N. F. Mott*, Dislocation and Mechanical Properties of Crystals. New York 1954
[6] *J. L. Lytton, L. A. Shepart, L. E. Dorn*, Trans. AIME (1958) 212
[7] *J. H. Bartlett, G. J. Dienes*, Phys. Rev. **89** (1953) 848
[8[ *A. C. Damask, G. J. Dienes, V. G. Weizer*, Phys. Rev. **113** (1959) 781

Isotopenpraxis 22 (1986) 3, S. 96 bis 99

# Bestimmung der Transfer-Koeffizienten für Strontium im System Boden—Pflanze

*T. Szabová*

(Institut für Radioökologie und Kerntechnikanwendung)[1])

*Es wurden Unterschiede zwischen den Transferkoeffizienten im System Boden—Pflanze für die vegetativen und die generativen Pflanzenorgane festgestellt. Bei den untersuchten Nutzpflanzen wurden keine Unterschiede zwischen den 2 applizierten Salzen des Strontiums gefunden — dem Nitrat und dem Chlorid. Bei der Bestimmung der Diskriminationsfaktoren wurden höhere Werte für die Knollen im Vergleich zu den Spinatblättern und den Erbsenkörnern bzw. zu den oberirdischen Teilen der Pflanzen ermittelt.*

*Differences between transfer coefficients of vegetative and generative plant organs in the soil-plant-system have been find out. No differences were found between the two salts of strontium — nitrate and chloride — applied to the useful plants which were investigated. The discrimination factors show higher values for the nodes in comparison to the leaves of spinach and to the pea pods, i.e. to the aboveground parts of the plant*

### Keywords

radionuclide migration; soils; strontium 90; strontium chlorides; strontium nitrates; vegetables

Die Radionuklidaufnahme von Pflanzen aus dem Boden ist durch die Transferkoeffizienten definiert. Diese Koeffizienten sind durch verschiedene Umweltparameter, wie das Klima, die Bodeneigenschaften, Bodentextur, die Kationenaustauscherkapazität, austauschbares Calcium, pH und Gehalt an organischer Substanz, beeinflußt. Die Koeffizienten variieren auch in Abhängigkeit von Pflanzenart, Entwicklungsstadium und untersuchtem Teil der Pflanze, sowie vom benutzten Isotop und seiner chemischen Applikationsform.

Bezüglich ihrer Bedeutung für die Strahlungsbelastung der Bevölkerung durch Verzehr von pflanzlicher bzw. tierischer Nahrung sind die Radionuklide $^{90}$Sr und $^{137}$Cs von größter Bedeutung.

Das Ziel dieser Arbeit war die Bestimmung der Transferkoeffizienten für stabiles Strontium in Abhängigkeit von der applizierten Menge und von der chemischen Form für einzelne Pflanzenarten.

[1]) Anschrift: ČSSR-040 61 Košice, Ústav rádioekológie a využitia jadrovej techniky, Komenského 9, P.O. Box A-41

### Material und Methoden

Benutzter Boden: schwere, lehm- und tonhaltige Braunerde, entnommen aus der oberen 20-cm-Schicht des Ackerbodens

Agrochemische Kennzeichnung des Bodens:

| | |
|---|---|
| pH/KCl | 4,61 |
| Humusprozent nach *Ťjurin* | 1,86 |
| austauschbares Calcium, mval/100 g | 10,40 |
| austauschbares Magnesium, mval/100 g | 2,75 |
| Summe der basisch wirkenden Kationen (S), mval/100 g | 15,70 |
| Gesamtsäurestärke ($H^+$), mval/100 g | 3,67 |
| gesamte Sorbtionskapazität (*T*), mval/100 g | 19,30 |

Grad der Sorptionssättigung, in % 81,04
auswaschbarer Teil des Strontiums, in ppm 12,80

Benutzte Pflanzenarten: Spinat, Erbse, Rettich

Die Analysen wurden mit Hilfe der Flammenemissionsspectroskopie durchgeführt.

Bodenproben von 6 kg wurden mit Handelsdünger homogenisiert und in die *Mitscherlich*-Gefäße eingefüllt. Nachdem die Samen ausgesät waren, wurde Strontium in Form von Nitrat oder Chlorid in Lösung zugesetzt. Die Früchte wurden im reifen Stadium geerntet. Die Transferkoeffizienten und die Diskriminationsfaktoren wurden für die Wurzel, den oberirdischen Teil und für die Früchte berechnet.

Für die Berechnung des Transferkoeffizienten wurde die Formel

$$TF_{BP} = \frac{\text{ppm Sr in Pflanzentrockenmasse}}{\text{ppm Sr in Trockenboden}}$$

verwendet, was einem Konzentrationsfaktor (CR) entspricht.

Der Diskriminationsfaktor ($D$) wurde nach der Formel

$$D = \frac{^{90}\text{Sr/Ca in der Pflanze}}{^{90}\text{Sr/Ca im Boden}}$$

berechnet.

## Ergebnisse

### *Rettich*

In den Experimenten wurde die Rettichsorte „ILKA" benutzt. Die Strontiumaufnahme wurde an einer ton- und lehmhaltigen Braunerde untersucht. Die Rettiche wurden in *Mitscherlich*-Gefäßen mit 6 kg Bodenvolumen angebaut. In jedes Gefäß wurden 12 Samen ausgesät, die nach ihrem Auskeimen zu 6 Pflanzen pro Gefäß selektiert wurden. Wir haben für jede Versuchsvariante vier Einzelversuche durchgeführt.
Nährstoff- und Strontiummengen:

15 g Ammoniumsulfat
4 g Superphosphat
4 g Kaliumsalz 60%

1. Kontrolle ohne Strontium
2. 10 mg Sr in der Form von $Sr(NO_3)_2$ in 1 kg Boden
3. 20 mg Sr in der Form von $Sr(NO_3)_2$ in 1 kg Boden
4. 100 mg Sr in der Form von $Sr(NO_3)_2$ in 1 kg Boden
5. 10 mg Sr in der Form von $SrCl_2$ in 1 kg Boden
6. 20 mg Sr in der Form von $SrCl_2$ in 1 kg Boden
7. 100 mg Sr in der Form von $SrCl_2$ in 1 kg Boden

Die Strontiumverbindung wurde in 1000 ml destilliertem Wasser gelöst und nach dem Aussäen dem Samen zugegeben. Während des Versuches wurden die Gefäße mit destilliertem Wasser begossen. Die Versuchsgefäße wurden in einem Glashaus aufgestellt.

Am Ende des Versuches wurden die unter- und oberirdischen Teile der Pflanzen in jedem Gefäß getrennt. Die Wurzeln wurden durch Waschen von den Bodenresten befreit. Die Blätter und die Knollen wurden bei 105 °C getrocknet.

Die Ergebnisse für Rettich sind in Tab. 1 zusammengestellt.

Tab. 1. Transferkoeffizienten ($TF_{BP}$) Boden – Pflanze für Rettich

| | Variante | Boden ppm Sr | oberird. Teil ppm Sr | $TF_{BP}$ | Knollen ppm Sr | $TF_{BP}$ |
|---|---|---|---|---|---|---|
| $Sr(NO_3)_2$ | Kontrolle | 8,12 | 29,3 | 3,6 | 13,8 | 1,69 |
| | 10 mg | 18,55 | 114,5 | 6,19 | 55,8 | 3,01 |
| | 20 mg | 32,65 | 169,5 | 5,19 | 101,1 | 3,09 |
| | 100 mg | 120,3 | 375,5 | 3,12 | 299,0 | 2,48 |
| $SrCl_2$ | 10 mg | 20,6 | 133,25 | 6,46 | 45,8 | 2,22 |
| | 20 mg | 29,0 | 135 | 4,65 | 64,0 | 2,20 |
| | 100 mg | 112,95 | 323,5 | 2,86 | 278,0 | 2,46 |

Die Anreicherung des Strontiums ist in den Blättern größer als in den Knollen. Die Transferkoeffizienten für Blätter und für Knollen sind kleiner für die Chlorid- als für die Nitratform des Strontiums. Bei einer Strontiummenge von 100 mg/1 kg Boden liegt der $TF_{BP}$-Wert für die Nitratform niedriger.

Bei der Chloridform wurden keine Unterschiede zwischen den Applikationsmengen des Strontiums festgestellt.

### *Erbse*

Der Versuchsboden — schwere, ton- und lehmhaltige Braunerde — wurde homogenisiert und für jedes Versuchsgefäß auf folgende Weise gedüngt:

5,5 g Superphosphat
6,0 g Calciumsalz
2,0 g Ammoniumsulfat

Für jede Versuchsvariante wurden vier Gefäße angesetzt. Nachdem die Gefäße mit gedüngtem Boden gefüllt waren, wurden in jedes Gefäß 6 Erbsenkörner gepflanzt. Dann wurden die Gefäße mit in 1000 ml destilliertem Wasser gelöstem Strontiumsalz begossen. Folgende Varianten wurden benutzt:

1. 10 mg $Sr(NO_3)_2$/1 kg Boden
2. 100 mg $Sr(NO_3)_2$/1 kg Boden
3. 10 mg $SrCl_2$/1 kg Boden
4. 100 mg $SrCl_2$/1 kg Boden.

Die Keimlinge wurden zu 4 Stück pro 1 Gefäß selektiert und bis zum Reifestadium belassen. Die Transferkoeffizienten wurden für Körner, Hülsen, den oberirdischen Teil und Wurzeln ausgewertet.

Die Ergebnisse für Erbsen sind in Tab. 2 zusammengefaßt.

Den niedrigsten Transferkoeffizienten haben wir in den Erbsenkörnern gefunden, und es konnten keine Unterschiede zwischen den Applikationsformen und Strontiummengen festgestellt werden.

Erhöhte Transferkoeffizienten wurden für die Hülsen ohne Körner ermittelt, wo auch Konzentrationsunterschiede gefunden wurden. Mit der steigenden Strontiumkonzentration nimmt hier der Transferkoeffizient ab. Es konnten keine Unterschiede zwischen den chemischen Formen des Strontiums gefunden werden.

Die höchste Transferkoeffizienten wurden für den oberirdischen Teil registriert (Stengel + Blätter). Es wurde kein Unterschied zwischen beiden Applikationsformen ermittelt; bei den höheren Sr-Mengen nimmt der Transferkoeffizient ab.

In den Wurzeln sind im Vergleich mit dem oberirdischen Teil die Transferkoeffizienten um etwa 1/3 niedriger, und es ist eine ähnliche Mengenabhängigkeit bemerkbar.

Tab. 2. Transferkoeffizienten für Erbsen

| | Variante | Boden ppm Sr | Körner ppm Sr | $TF_{BP}$ | Hülsen ohne Körner ppm Sr | $TF_{BP}$ | Oberird. Teil ppm Sr | $TF_{BP}$ | Wurzel ppm Sr | $TF_{BP}$ |
|---|---|---|---|---|---|---|---|---|---|---|
| | Kontrolle | 12,67 | 1,18 | 0,09 | 39,5 | 3,11 | 88,7 | 7,0 | 58,0 | 4,4 |
| $Sr(NO_3)_2$ | 10 mg | 19,87 | 1,82 | 0,09 | 64,4 | 3,24 | 132,5 | 6,66 | 80,4 | 4,04 |
| | 100 mg | 95,02 | 10,5 | 0,11 | 225,0 | 2,36 | 513,0 | 5,39 | 303,0 | 3,18 |
| $SrCl_2$ | 10 mg | 21,82 | 2,18 | 0,09 | 68,3 | 3,13 | 133,0 | 6,09 | 90,5 | 4,14 |
| | 100 mg | 123,25 | 10,9 | 0,09 | 325,5 | 2,6 | 507,0 | 5,11 | 348,0 | 2,82 |

### *Spinat*

Spinat haben wir in Gefäßen mit 11 kg Bodenvolumen im Frühbeet angebaut. Nach der Düngung mit anorganischen Düngemitteln (10 g Superphosphat, 12 g Kaliumsalz, 15 g Ammoniumsulfat) wurde der Boden in die Frühbeetgefäße gefüllt. Die Samen wurden in 2 Reihen mit je 30 Samen ausgesät. Das Strontium wurde als Chlorid und Nitrat appliziert, und zwar:

10 mg Strontium pro 1 kg Boden und

100 mg Strontium pro 1 kg Boden in gelöster Form.

Bei der Probennahme wurden die Blätter und die Wurzel voneinander getrennt und die Transferkoeffizienten für die beiden Pflanzenteile getrennt berechnet.

Die ermittelten Ergebnisse sind in Tab. 3 zusammengestellt.

Tab. 3. Transferkoeffizienten für Spinat

| | Variante | Boden ppm Sr | Blätter ppm Sr | $TF_{BP}$ | Wurzeln ppm Sr | $TF_{BP}$ |
|---|---|---|---|---|---|---|
| | Kontrolle | 12,4 | 42,3 | 3,41 | 41,15 | 3,31 |
| $Sr(NO_3)_2$ | 10 mg | 21,3 | 62,35 | 2,92 | 65,7 | 3,08 |
| | 100 mg | 123,0 | 214,0 | 1,73 | 220,0 | 1,78 |
| $SrCl_2$ | 10 mg | 17,4 | 62,05 | 3,56 | 59,8 | 3,43 |
| | 100 mg | 120,0 | 208,0 | 1,73 | 172,0 | 1,43 |

## Diskussion

Aus den Versuchen ergibt sich ein Unterschied zwischen den Transferkoeffizienten in vegetativen und generativen Organen. Die vegetativen Organe weisen höhere Transferkoeffizienten als die generativen auf.

Bei den untersuchten Nutzpflanzen zeigten sich keine Unterschiede in den Transferkoeffizienten für die Nitrat- und Chloridformen des Strontiums. Eine Erhöhung der Applikationsmenge des Strontiums bewirkte bei beiden Salzen eine Erniedrigung der Transferkoeffizienten.

Ein Vergleich der gefundenen Ergebnisse mit Literaturangaben [1, 2], wo ein Wert für Pflanzen von $1{,}6 \cdot 10^{-3}$ bis 9,3 angegeben wird, ergibt eine gute Übereinstimmung.

Ein Vergleich der erhaltenen Werte mit Angaben von *Ng* et al. [3] ergibt folgendes Bild: Diese Autoren geben die Transferkoeffizienten ($B_v$) und die Konzentrationsfaktoren (*CR*) für die einzelne Isotope, Pflanzenarten und Bodentypen bei verschiedenen Versuchsbedingungen an.

| Stabiles Strontium | unsere Ergebnisse | Literaturangaben | |
|---|---|---|---|
| Rettich | 2,22–3,09 | 3,07–3,3 | [4] |
| Erbsenkörner | $8 \cdot 10^{-2}$–$1{,}1 \cdot 10^{-1}$ | $3{,}9 \cdot 10^{-1}$ | [5] |
| Spinat | 1,73–3,56 | 3,69 $-1{,}43 \cdot 10^{-1}$ | [6] |

Unsere Ergebnisse haben einen vorläufigen Charakter, und es sind weitere experimentelle Überprüfungen an einer größeren Zahl von Bodentypen und Pflanzenarten notwendig.

Bei dem Strontiumtransfer in die Pflanzen kommt es auch zu einer Diskrimination von Radiostrontium in bezug auf Calcium, wobei der Diskriminationsfaktor meistens kleiner als 1 ist. Der Diskriminationsfaktor in den Erbsenkörner ist außerdem kleiner als im oberirdischen Teil der Pflanze und zugleich wesentlich kleiner als in Rettich und Spinat.

Die Diskriminationsfaktoren für Rettichknollen, Spinatblätterund Erbsenkörner sind in Tab. 4 zu sehen.

Eingegangen am 23. 4. 1985
Überarbeitete Fassung am 30. 5. 1985

## Literatur

[1] U.S. Nuclear Regulatory Commision, Regulatory Guide 1.109, Calculation of Annual Doses to Man from Routine Releases of Reactor Effluents for the Purpose of Evaluating Compliance with 10 CFR Part 50, Appendix I., October 1977, Revision 1

[2] *Y. C. Ng*, Nucl. Safety 23 (1982) 57

Tab. 4. Diskriminationsfaktorwerte für einzelne Nutzpflanzen

| Variante | | Rettichknollen $TF_{BP}$ | Rettichknollen $D$ | Spinatblätter $TF_{BP}$ | Spinatblätter $D$ | Erbsenkörner $TF_{BP}$ | Erbsenkörner $D$ | Erbse – oberird. Teil $TF_{BP}$ | Erbse – oberird. Teil $D$ |
|---|---|---|---|---|---|---|---|---|---|
| Kontrolle | | 1,7 | 0,975 | 3,41 | 1,175 | 0,09 | 0,464 | 7,0 | 0,809 |
| $Sr(NO_3)_2$ | 10 mg | 3,01 | 1,69 | 2,92 | 0,647 | 0,09 | 0,259 | 6,66 | 0,741 |
| | 20 mg | 3,09 | 1,92 | | | | | | |
| | 100 mg | 2,48 | 1,8 | 1,73 | 0,379 | 0,11 | 0,54 | 5,39 | 0,582 |
| $SrCl_2$ | 10 mg | 2,22 | 1,42 | 3,56 | 1,048 | 0,09 | 0,503 | 6,09 | 0,724 |
| | 20 mg | 2,2 | 1,475 | | | | | | |
| | 100 mg | 2,46 | 1,873 | 1,73 | 0,461 | 0,08 | 0,409 | 5,11 | 0,623 |

$TF_{BP}$ = Transferkoeffizient
$D$ = Diskriminationsfaktor

[3] *Y. C. Ng, C. S. Colsher, S. E. Thompson*: Soil – to – Plant Concentration Factors for Radiological Assessments, NUREG/CR-2975, UCID-19463 (Lawrence Livermore Laboratory, Livermore, Cal.), November 1982

[4] *J. W. Neel, J. H. Olafson, A. J. Steen, B. E. Gillooly, H. Nishita, K. H. Larson*, Soil – Plant Interrelationships with Respect to the Uptake of Fission Products: I. The Uptake of $^{90}$Sr, $^{137}$Cs, $^{106}$Ru, $^{144}$Ce and $^{91}$Y. UCLA-247, 1953, University of California at Los Angels, West Los Angels, Cal.

[5] *E. P. Hardy, B. G. Bennett, L. T. Alexander*, Radionuclide Uptake by Cultivated Crops, HASL-321: I-19-37. ERDA Health and Safety Laboratory, New York 1977

[6] *L. Fredricksson, H. Lönsjö, A. Eriksson*, Studies on Plant Accumulation of Fission Products under Swedish Conditions XII. Absorption of $^{90}$Sr and $^{137}$Cs from Soil by Vegetable Crops. C-4387-28, Research Institute of National Defense, Dept. 4, Stockholm, Sweden, 1969

Isotopenpraxis 22 (1986) 3, S. 99 bis 105

# Radionuklidbestimmungen in Umweltmedien zur Ermittlung der Strahlenbelastung der Bevölkerung der DDR*)

*E. Ettenhuber, B. Kiehl, D. Obrikat, H.-U. Siebert, J. Thiele*

(Staatliches Amt für Atomsicherheit u. Strahlenschutz der DDR, Institut für Umwelt-Überwachung)[1]

*Ausgehend von den Quellen der zivilisationsbedingten Strahlenbelastung der Bevölkerung der DDR und den daraus abgeleiteten Umweltüberwachungsprogrammen werden anhand der Umweltüberwachung für eine Kernanlage die strahlungsbelastungsrelevanten Meßgrößen und die entsprechenden meßtechnischen Maßnahmen dargestellt. Es werden eine Reihe von Methoden zur Probenaufbereitung, Radionuklidanreicherung, chemischen Trennung und Präparation sowie zur Konzentrationsbestimmung von Radionukliden und stabilen Elementen in Proben aus Umweltüberwachungsprogrammen, insbesondere unter Verwendung von Niederaktivitätsmeßverfahren, beschrieben.*
*Mit Hilfe der Ergebnisse von Radionuklidbestimmungen in Umweltproben werden auf der Grundlage von Strahlenbelastungsmodellen die Beiträge der zivilisationsbedingten Strahlenbelastung aus der Umwelt zur Gesamtstrahlenbelastung der Bevölkerung der DDR abgeschätzt.*

*Starting from the sources of man-made radiation exposure of the population of the GDR and the environmental monitoring programmes derived therefrom, measured values, relevant to radiation exposure, and essential measuring activities are presented, using the environmental monitoring of a nuclear facility as an example.*
*A number of methods are described which are employed for sample processing, radionuclide enrichment, chemical separation and preparation as well as for measurement of radionuclide and stable element concentrations in samples received from environmental monitoring programmes with especial regard to the use of low-level activity measuring methods. Using results from radionuclide determinations in environmental samples, the contributions of man-made radiation exposure from the environment to the radiation exposure of the GDR population are estimated on the basis of exposure calculation models.*

## Keywords

alpha spectroscopy; atomic absorption spectroscopy; beta spectroscopy; counting techniques; environment; gamma spectroscopy; quantitative chemical analysis; radiation doses; radiation monitoring; radioassay; sample preparation

## 1. Nukleare Umweltüberwachungsprogramme der DDR

Das Ziel des Strahlenschutzes besteht in der Gewährleistung des Schutzes des Lebens und der Gesundheit des Menschen vor der schädigenden Einwirkung ionisierender Strahlung [1]. Eine der wichtigsten Konsequenzen daraus ist, den Anteil der zusätzlich zur natürlichen Strahlenbelastung auftretenden zivilisationsbedingten Strahlenbelastung mit gesellschaftlich zu rechtfertigendem Aufwand auf ein von der Gesellschaft akzeptiertes sehr niedriges Niveau zu begrenzen. Das erfordert eine klare Abgrenzung der zivilisationsbedingten und der natürlichen Strahlenbelastung, wobei die Strahlenbelastungen, die durch die Tätigkeit des Menschen hervorgerufen bzw. verändert werden, als zivilisationsbedingte und die ohne den Einfluß

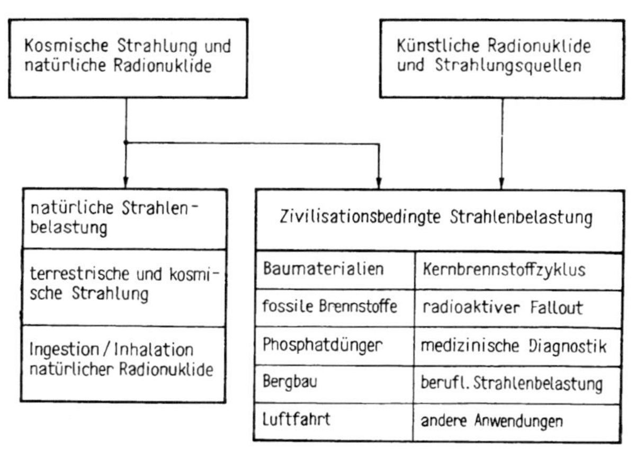


Abb. 1. Komponenten der natürlichen und zivilisationsbedingten Strahlenbelastung

*) Vortrag, gehalten auf dem 17. Internationalen Symposium „Strahlenschutzphysik" im März 1985 in Dresden

[1]) Anschrift: DDR-1157 Berlin, Waldowallee 117

des Menschen zustande kommenden als natürliche Strahlenbelastungen definiert werden (Abb. 1).

Die primordialen Radionuklide, ihre radioaktiven Zerfallsprodukte und die kosmische Strahlung verursachen eine aus einer terrestrischen und kosmischen Komponente bestehende äußere und über die Inhalation und Ingestion natürlicher Radionuklide eine innere Strahlenbelastung. Durch die seine Umwelt verändernde Tätigkeit des Menschen ergibt sich in verschiedenster Weise, so durch den Einsatz mineralischer Baumaterialien, fossiler Brennstoffe und Phosphatdünger, durch den Bergbau und die Luftfahrt, eine zivilisationsbedingte Strahlenbelastung aus diesen natürlichen Quellen. Eindeutig der zivilisationsbedingten Strahlenbelastung zuzuordnen sind die bei beruflicher oder medizinischer Anwendung von Radionukliden oder anderen Strahlenquellen, die durch den radioaktiven Fallout aus den Kernwaffenversuchen und die aus dem Kernbrennstoffzyklus resultierenden Strahlenbelastungen.

Das Staatliche Amt für Atomsicherheit und Strahlenschutz der DDR veranlaßt oder führt selbst auf alle Komponenten der zivilisationsbedingten Strahlenbelastung gerichtete Umweltüberwachungsprogramme durch, um die Kontamination der Umweltmedien und die dadurch oder durch die Veränderung der Expositionsbedingungen zustande kommenden zivilisationsbedingten Strahlenbelastungen zu ermitteln und eine Einschätzung der Strahlenbelastung der Bevölkerung der DDR durchführen zu können. Diese Überwachungsprogramme dienen zum Nachweis der Einhaltung gesetzlicher Forderungen [1] oder haben den Charakter von Erhebungsuntersuchungen. Sie werden

- auf dem gesamten Territorium zur Ermittlung der Auswirkungen des globalen Fallouts und der Kontamination der Baumaterialien durch natürliche Radionuklide,
- in der Umgebung von Kernanlagen,
- im Bergbaugebiet,
- in der Umgebung konventioneller Kraftwerke sowie
- in der Ostsee und den Ostseerandgewässern

durchgeführt.

## 2. Strahlenbelastungsrelevante Meßgrößen und meßtechnische Maßnahmen

Die in den einzelnen Umweltüberwachungsprogrammen zur Anwendung kommenden meßtechnischen Maßnahmen sind unterschiedlich und müssen sich an den für die Ermittlung der Strahlenbelastung notwendigen Meßgrößen orientieren.

Abb. 2 zeigt als Beispiel die möglichen Belastungswege in der Umwelt von Kernanlagen, die bei gas- bzw. aerosolförmigen Auswürfen in die Atmosphäre und bei flüssigen Auswürfen in die Oberflächengewässer zu betrachten sind. Die in die Atmosphäre freigesetzten Radionuklide führen durch die Submersion und die Bodenstrahlung zu einer äußeren Strahlenbelastung, bei ihrer Inhalation zu einer inneren Strahlenbelastung sowie nach Ablagerung auf der Erdoberfläche über ökologische Ketten zu einer Nahrungsmittelkontamination und somit zu einer inneren Strahlenbelastung durch Ingestion. Durch die flüssigen Auswürfe wird ebenfalls eine äußere Strahlenbelastung, vor allem aber über ökologische Ketten eine innere Strahlenbelastung durch Ingestion hervorgerufen. In den internationalen Empfehlungen [2] wird darauf orientiert, zur Bestimmung

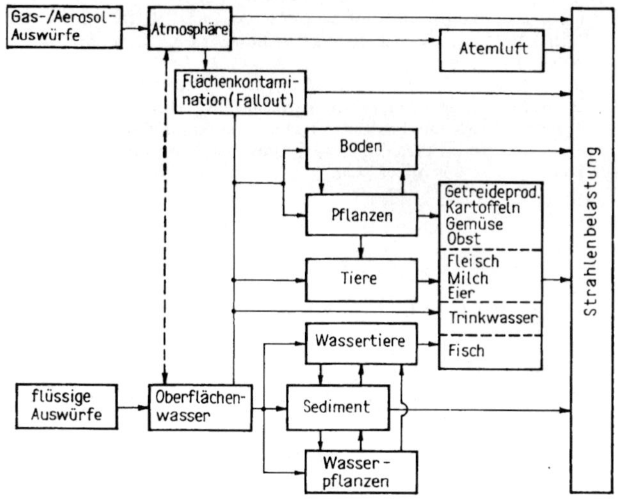


Abb. 2. Schema der Strahlenbelastungswege durch radioaktive Auswürfe aus Kernanlagen

der Strahlenbelastung möglichst die Endstufen der Belastungswege zu überwachen. In vielen Fällen ist es unmöglich, die Endstufen der Belastungswege direkt zu überwachen. Dann ist es unumgänglich, andere Stufen der Belastungswege zur Überwachung zu nutzen, aber auch in allen übrigen Fällen können, wenn die notwendigen Belastungsmodelle vorhanden sind, zur Optimierung des Aufwandes geeignete Stufen der Belastungswege überwacht werden. Aus Abb. 2 wird deutlich, daß die bodennahe Atmosphäre und die Oberflächengewässer die zentralen Stufen der Belastungswege darstellen. Von gleicher Bedeutung sind innerhalb der aquatischen Belastungswege die Sedimente, die als Radionuklidlangzeitreservoir eine Kontaminationsquelle für das übrige aquatische Ökosystem bilden können. Während die Überwachung der Oberflächengewässer und der Sedimente mit geringem Aufwand möglich ist, erfordert die Überwachung der bodennahen Atmosphäre wesentlich höheren Aufwand. Sind jedoch die leicht zu messende äußere Strahlenbelastung, die auf Grund der Kontamination der Atmosphäre und der Flächenkontamination zustande kommt, und parallel dazu die ebenfalls leicht zu messende Flächenkontamination bekannt, dann läßt sich bei Kenntnis des freigesetzten Radionuklidspektrums die Radionuklidkonzentration in der Atmosphäre mit ausreichender Genauigkeit abschätzen. Für die übrigen Biomedien kann die Kontamination aus der Flächenkontamination berechnet und durch Stichproben kontrolliert werden. Damit ergibt sich ein System meßtechnischer Maßnahmen der Umweltüberwachung [3], das aus den Komponenten

- Messung der Aktivitätsauswürfe und der Auswurfbedingungen,
- Messung der Ortsdosis bzw. Ortsdosisleistung sowie
- Messung der Flächenkontamination und der Radionuklidkonzentration in Oberflächengewässern und Biomedien durch probenanalytische Bestimmungen sowie

besteht.

Die probenanalytischen Bestimmungen von Radionuklidkonzentrationen in Umweltmedien spielen in diesem System eine wichtige Rolle.

## 3. Anforderungen an probenanalytische Bestimmungsmethoden

Im Ergebnis der Analyse der Belastungswege resultiert aus den Überwachungsprogrammen eine große Zahl verschiedener Probenmedien, in denen bis zu 50 für den Strahlenschutz relevante Radionuklide und Bezugselemente zu bestimmen sind. Die Hauptprobenmedien sind dabei Luft der bodennahen Atmosphäre, Wasser, Boden, Sedimente, Pflanzen und Nahrungsmittel. Tab. 1 zeigt die große Variationsbreite der Probenmedien, der Aktivitätskonzentration und der zu bestimmenden Radionuklide für die einzelnen Überwachungsprogramme.

In den zu analysierenden Proben sind in den meisten Fällen Aktivitätskonzentrationen $< 1$ Bq/g und Konzentrationen stabiler Elemente im $\mu$g/g-Bereich zu bestimmen, so daß für die Analyse der Einsatz von Verfahren der Niederaktivitätsmeßtechnik bzw. der Spurenanalyse erforderlich ist. Da in vielen Fällen die Aktivitätskonzentrationen so gering sind, daß sie durch ein Strahlungsmeß- bzw. Analysenverfahren direkt nicht bestimmbar sind, ergibt sich die Notwendigkeit, vor der Analyse chemische nuklidspezifische und nuklidunspezifische Anreicherungsverfahren durchzuführen. Die Wahl des geeigneten Anreicherungsverfahrens und Meßverfahrens ist abhängig von

- dem Probenmedium und der Probenmasse,
- den chemischen und physikalischen Eigenschaften der zu bestimmenden Radionuklide,
- den Strahlungseigenschaften der zu bestimmenden Radionuklide,
- der erforderlichen unteren Nachweisgrenze bzw. dem tatsächlichen Aktivitätsniveau der Probe und
- der erforderlichen Analysengenauigkeit.

Die zur Realisierung der probenanalytischen Bestimmungen bereitzustellenden Meßverfahren müssen beim gegenwärtigen Umfang der Umweltüberwachungsprogramme in der DDR einen Durchsatz von etwa 3000 Proben pro Jahr mit etwa 15000 Radionuklid- bzw. Elementbestimmungen gestatten. Das erfordert eine Zentralisierung der aufwendigen Probenbearbeitung und Meßtechnik sowie ein hohes Niveau der Meßautomatisierung, der Datenerfassung, -verarbeitung und -dokumentation [4–6].

## 4. Methoden der Niederaktivitäts- und Spurenelementbestimmung

Die Bestimmung der Aktivitätskonzentration von Radionukliden und der Spurenelementkonzentration in Umweltproben erfordert ein System

- der Probenaufbereitung und nuklidunspezifischen bzw. nuklidspezifischen Anreicherung,
- der chemischen Methoden zur Abtrennung und Isolierung der zu bestimmenden Radionuklide sowie
- der Niederaktivitäts- und Analysenmeßtechnik.

### 4.1. *Probenaufbereitung und Radionuklidanreicherung*

In einem großen Teil der Proben aus den Überwachungsprogrammen ist die direkte Bestimmung der Radionuklidkonzentrationen nicht möglich, da infolge der geringen Aktivitätskonzentration Probenmassen zur Bestimmung eingesetzt werden müssen, die analytisch und meßtechnisch durch den hohen Anteil an Ballaststoffen kaum oder nicht mehr handhabbar sind. Deshalb müssen der Aktivitätsbestimmung Probenaufbereitungsschritte zur meist mechanischen Umwandlung der Rohprobe in eine zur Analysendurchführung effektivere Form und Konsistenz sowie Anreicherungsschritte zur größtmöglichen Massen- bzw. Volumenreduzierung der Probe durch weitestgehende Entfernung der überwiegend organischen Probenmatrix vorangestellt werden.

Abb. 3 gibt einen Überblick über die angewandten Probenaufbereitungs und Anreicherungsverfahren. In der Regel ist die Trockenveraschung bei maximal 450 °C das einfachste und zumeist effektivste Verfahren zur Entfernung der organischen Probenmatrix und damit der nuklidunspezifischen Anreicherung.

Eine Sonderform der Veraschung, insbesondere für Proben mit geringer Einsatzmenge, stellt die Tieftemperaturveraschung [7, 8] zur Vermeidung von Verlusten an besonders leichtflüchtigen Elementen dar. Zur Umgehung dieser apparativ aufwendigen Methode kann auch der Naßaufschluß der vorgetrockneten und zerkleinerten Probe erfolgen. Dieses Verfahren, das damit gleichzeitig Ausgangspunkt der anzuwendenden Trenn- und Präparationsver-

Tab. 1. Charakteristika von Umweltüberwachungsproben

| | Radionuklide | Probenart | Aktivitätskonzentration |
|---|---|---|---|
| Kernanlagen | $^{3}$H, $^{14}$C, $^{54}$Mn,<br>$^{58}$Co, $^{60}$Co, $^{90}$Sr,<br>$^{95}$Zr/Nb, $^{110m}$Ag,<br>$^{131}$J, $^{134}$Cs, $^{137}$Cs,<br>$^{152}$/$^{154}$Eu | Fallout<br>Boden, Gras<br>Luftfilter<br>Oberflächenwässer<br>Abwässer | $<$ 100 Bq/m$^2$<br>$<$ 1000 mBq/g<br>$<$ 100 mBq/m$^3$<br>$<$ 5 Bq/l<br>$<$ 100 Bq/l |
| globaler Fallout | $^{3}$H, $^{14}$C, $^{90}$Sr,<br>$^{95}$Zr/Nb, $^{103}$Ru,<br>$^{137}$Cs, $^{144}$Ce | Luft<br>Fallout<br>Boden, Gras<br>Regenwasser<br>Milch | 1 – 500 mBq/m$^3$<br>0·1 – 0·5 Bq/m$^2$<br>1 – 10 mBq/g<br>0·001 – 0·01 Bq/l<br>0·01 – 0·5 Bq/l |
| Bergbaugebiete | $^{40}$K, $^{210}$Pb, $^{210}$Po,<br>$^{224}$Ra, $^{226}$Ra,<br>Uranium, $^{230}$Th | Oberflächenwasser<br>Gestein<br>Baumaterial | 0·01 – 50 Bq/l<br>1 – 50000 Bmq/g<br>5 – 500 Bmq/g |
| konventionelle Kraftwerke | $^{40}$K, $^{210}$Pb, $^{210}$Po, $^{224}$Ra,<br>$^{226}$Ra, $^{230}$Th, $^{232}$Th | Asche<br>Kohle | 10 – 1000 mBq/g<br>1 – 100 mBq/g |
| Ostsee | $^{3}$H, $^{90}$Sr, $^{137}$Cs,<br>$^{210}$Pb, $^{210}$Po,<br>Spurenelemente | Wasser<br>Fisch<br>Sedimente | 0·001 – 0·05 Bq/l<br>0·01 – 5 mBq/g<br>1 – 500 mBq/g |

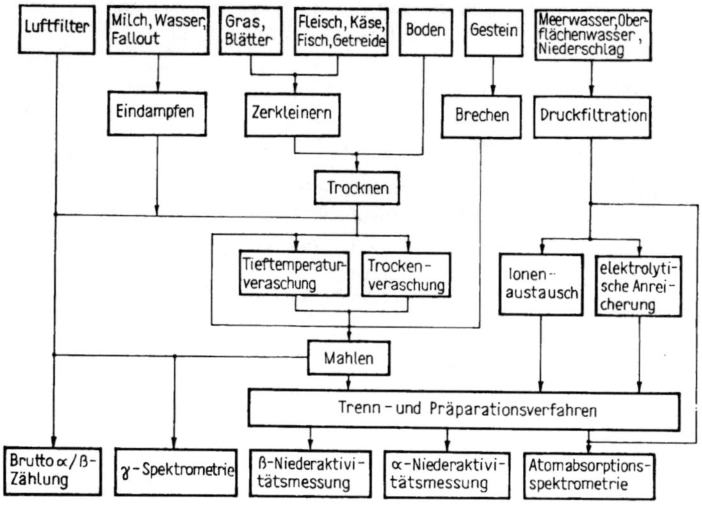


Abb. 3. Schema der Aufbereitung und der analytischen Bestimmungsmethode für Umweltproben

fahren ist, eignet sich jedoch mehr zur Radionuklid- als zur Spurenelementanalytik, da dafür relativ große Säuremengen eingesetzt werden.

Vor der Veraschung sind in Abhängigkeit von der Probenart verschiedene Aufbereitungsschritte wie Filtration, Verdampfung, Trocknung und Zerkleinerung erforderlich, um den Veraschungsprozeß effektiv zu gestalten. Nach der Veraschung ist in fast allen Fällen eine Mischung und Homogenisierung der Probenrückstände erforderlich. Die so erzielten Anreicherungsfaktoren betragen ca. 10—100 für Biomedien und Nahrungsmittel und bis zu 1000 für wäßrige Proben. In einzelnen Fällen, insbesondere bei großvolumigen wäßrigen Proben, ist eine nuklidspezifische Anreicherung von Radionukliden möglich, indem sie unter entsprechenden Reaktionsbedingungen mit einem nur für ein Element oder eine Elementgruppe spezifisch wirkenden Ionenaustauscher behandelt werden. Das trifft zu für die Anreicherung von $^{137}$Cs aus See- und Oberflächenwasserproben mit einem Volumen bis zu 100 l an Triammoniumdodekamolybdatophosphat (AMP) [9]. Hierbei können Anreicherungsfaktoren bis zu 1000 erreicht werden. Eine andere Form der nuklidspezifischen Anreicherung ist für die Anreicherung von $^{3}$H aus Oberflächen- und Meerwasserproben die elektrolytische Zersetzung [10]. Dabei werden Anreicherungsfaktoren von 10—20 erreicht.

Nach den Probenaufbereitungsschritten ist zu entscheiden, ob an den Probenrückständen eine direkte integrale oder spektrometrische Aktivitätsmessung durchgeführt werden kann oder ob vorher analytisch-chemische Trennoperationen zur Entfernung der die Bestimmungsmethode störenden Elemente und Radionuklide erforderlich sind.

### *4.2. Trenn- und Präparationsverfahren*

Obwohl eine Reihe hochempfindlicher und selektiver Meßverfahren zur direkten Radionuklidkonzentrationsbestimmung in Umweltproben zur Verfügung steht, sind Trenn- und Präparationsverfahren in vielen Fällen unumgänglich. Dabei wird auch ein weiterer Anreicherungseffekt von einigen Gramm Glührückstand der Probe auf wenige Milligramm bzw. Mikrogramm des Meßpräparates erreicht. Erster Schritt der analytischen Bearbeitung der Veraschungsrückstände der Proben ist in der Regel ihr Aufschluß. In Abhängigkeit vom Probenmedium, der erforderlichen Probenmenge und den zu bestimmenden Radionukliden werden saure Aufschlüsse, Säureauszüge und alkalische Schmelzen durchgeführt.

Zur Isolierung der zu bestimmenden Radionuklide von anderen Radionukliden und störenden Matrixelementen werden verschiedene Trennverfahren eingesetzt:

— Fällungen und Mitfällungen werden überwiegend am Anfang analytisch-chemischer Trennungsgänge zur Abtrennung von störenden Spuren- bzw. Matrixelementen oder als Gruppentrennung vor einer nachfolgenden Elementtrennung durchgeführt. Typische und effektive Beispiele hierfür sind in den angewandten Bestimmungsverfahren die $BaSO_4$-Fällung bei der Radium-Bestimmung [11], die $Fe(OH)_3$-Fällung bei der $^{137}$Cs-Bestimmung [12] und die $Ca(COO)_2$-Fällung bei der Thoriumbestimmung [13].
— Extraktion und Ionenaustausch werden überwiegend zur selektiven Abtrennung der zu bestimmenden Radionuklide eingesetzt. Für die Bestimmung von Transuranen und Thorium werden ternäre und quaternäre Alkylammoniumverbindungen als Extraktionsmittel verwendet [14]. Der Einsatz von Di-(2-ethyl-hexyl)-phosphorsäure (HDEHP) erfolgt im wesentlichen bei der Bestimmung einiger Spaltproduktnukilide (z.B. $^{90}$Sr) [15, 16]. Als spezieller Ionenaustauscher wird Triammoniumdodekamolybdatophosphat (AMP) zur selektiven Abtrennung von $^{137}$Cs eingesetzt [12].

Zur radiometrischen Messung der isolierten Radionuklide mit α-/β-Niederaktivitätsmeßverfahren ist in fast allen Fällen die Herstellung eines Meßpräparates mit möglichst geringer Flächenmasse erforderlich. Dabei werden:

— Fällungen der zu bestimmenden Radionuklide zusammen mit einer inaktiven Trägersubstanz zur Meßpräparateherstellung insbesondere von β-strahlenden Nukliden ($^{89}$Sr, $^{90}$Sr, $^{137}$Cs) eingesetzt [12, 15, 17];
— elektrolytische Abscheidungen der zu bestimmenden Radionuklide zur Meßpräparateherstellung vorwiegend bei α-strahlenden Nukliden eingesetzt, um möglichst „massefreie" Präparate zur Gewährleistung einer hohen Meßeffektivität und einer guten spektrometrischen Auflösung zu erhalten. Dabei wird sowohl die spontane elektrolytische Abscheidung (z.B. $^{210}$Pb/$^{210}$Bi/$^{210}$Po-Bestimmung [18]) als auch die Elektrolyse aus anorganischen Lösungen bei kleinen Spannungen für die Her-

Tab. 2. Chemisch-analytische Trenn- und Bestimmungsverfahren

| Radionuklide | Trennverfahren und Bestimmungsmethoden | Probenart | Probenmenge | untere Nachweisgrenze |
|---|---|---|---|---|
| $^{230}$Th, $^{232}$Th | Th-Mitfällung an Calciumoxalat/Th-Extraktion/elektrolytische Abscheidung/α-Spektrometrie | Wasser<br>Pflanzen<br>Boden, Gestein | 1 l<br>100 g<br>0,5 g | 1 mBq/l<br>0,01 mBq/g<br>2 mBq/g |
| $^{224}$Ra | Ra-Mitfällung an $BaSO_4$/α-Zählung/$^{222}$Rn-Emanation/α-Zählung/Subtraktion der $^{226}$Ra-Aktivität von der Gesamtaktivität | Wasser<br>Pflanzen<br>Boden | 1 l<br>100 g<br>5 g | 10 Bmq/l<br>0,1 Bmq/g<br>2 mBq/g |
| $^{226}$Ra | Ra-Mitfällung an $BaSO_4$/$^{222}$Rn-Emanation/α-Zählung | Wasser<br>Pflanzen<br>Boden | 1 l<br>100 g<br>5 g | 10 mBq/l<br>0,1 Bmq/g<br>2 mBq/g |
| $^{14}$C | $BaCO_3$-Fällung/Flüssigkeitsszintillationsspektrometrie | Luft | 10 m³ | 1 mBq/m³ |
| $^{89}$Sr | $SrCO_3$-Fällung/β-Zählung/Subtraktion der $^{90}$Sr-Aktivität von der Gesamtaktivität | Wasser<br>Boden, Pflanzen | 25 l<br>500 g | 1 mBq/l<br>50 mBq/g |
| Uranium | Extraktion mit Essigsäureethylester/Messung der Fluoreszenzausbeute | Wasser<br>Pflanzen<br>Boden | 0,2 ml<br>100 g<br>1 g | 0,005 µg/ml<br>0,01 µg/g<br>1 µg/g |
| $^{90}$Sr | $^{90}$Y-Extraktion/Yttriumoxalatfällung/β-Zählung | Wasser<br>Boden, Pflanzen | 25 l<br>500 g | 0,5 mBq/l<br>0,04 mBq/g |
| $^{239}$Pu | Extraktion mit Alamin/elektrolytische Abscheidung/α-Spektrometrie | Wasser<br>Pflanzen | 1 l<br>1 kg | 0,6 mBq/l<br>0,0006 mBq/g |
| $^{137}$Cs | Ionenaustausch/Hexachloroplatinatfällung/β-Zählung | Wasser<br>Pflanzen<br>Boden | 100 l<br>0,5 kg<br>30 g | 0,1 mBq/l<br>0,06 mBq/g<br>0,5 mBq/g |
| $^{210}$Pb, $^{210}$Bi, $^{210}$Po | spontane elektrolytische Abscheidung auf Ni/β-Zählung/α-Zählung | Wasser<br>Pflanzen<br>Boden | 1 l<br>10 g<br>2 g | 20 mBq/l<br>2 mBq/g<br>10 mBq/g |
| $^{3}$H | elektrolytische Anreicherung/Flüssigkeitsszintillationsspektrometrie | Wasser<br>Luft | 4 ml<br>10 m³ | 1 Bq/l<br>5 mBq/m³ |

stellung von Transuran- und Thoriumpräparaten angewendet [14].

Eine Zusammenstellung der bei den einzelnen Radionuklidbestimmungen angewandten Trenn- und Präparationsverfahren ist in Tab. 2 enthalten [19].

Bei chemischen Analysenmethoden mit mehreren Trennschritten können auftretende Radionuklidverluste durch Ausbeutebestimmungen nach Zusatz von inaktivem Trägermaterial oder radioaktiven Tracern korrigiert werden. Die notwendigen Ausbeutebestimmungen erfolgen mit Hilfe der Atomabsorptionsspektrometrie, gravimetrisch, titrimetrisch oder durch Aktivitätsmessung.

### 4.3. *Konzentrationsbestimmung von Radionukliden und stabilen Elementen*

Abb. 4 gibt einen Überblick über die zur Bestimmung von Radionukliden und stabilen Elementen eingesetzten Meßverfahren.

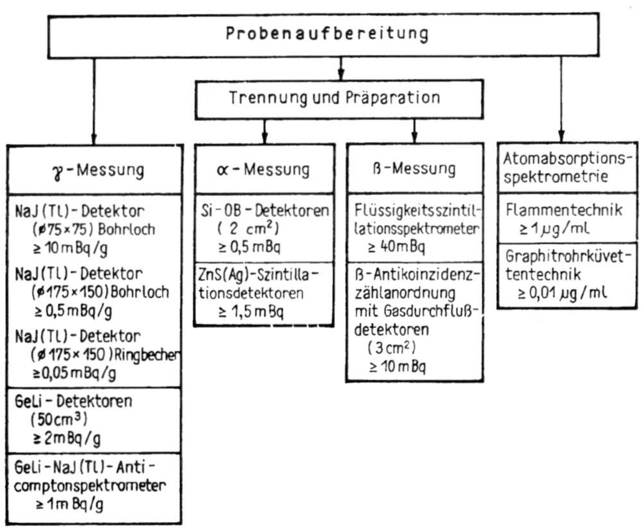


Abb. 4. Methoden der Aktivitäts- und Spurenelementbestimmung in Umweltproben

*4.3.1. Bestimmung von α-strahlenden Nukliden.* Für die integrale α-Aktivitätsmessungen werden ZnS(Ag)-Szintillationsdetektoren eingesetzt. Voraussetzung für eine Radionuklidkonzentrationsbestimmung ist dabei das Vorliegen nuklidreiner Präparate als Ergebnis chemischer Trenn- und Präparationsverfahren (z.B. $^{224}$Ra-, $^{226}$Ra- und $^{210}$Po-Bestimmung). Die untere Nachweisgrenze beträgt für die eingesetzten Meßanordnungen bei 100 min Meßzeit 1,5 mBq. Dabei wird als untere Nachweisgrenze der Meßwert angegeben, der mit einem $2\sigma$-Fehler von $\pm 100\%$ vom jeweiligen Nulleffekt unterschieden werden kann.

Die Verwendung von Si-Halbleiterspektrometern erlaubt mit hoher Selektivität und geringem Nulleffekt einen empfindlicheren Nachweis mehrerer gleichzeitig vorliegender α-strahlender Nuklide. Dabei werden an die für α-spektrometrische Messungen benötigten Meßpräparate besonders hohe Anforderungen bezüglich Homogenität und „Massefreiheit" gestellt. Bei der Aktivitätsbestimmung von $^{230}$Th, $^{232}$Th, $^{238}$U und $^{239}$Pu wird bei einer Meßzeit von 1000 min mit Si-OB-Detektoren mit einer effektiven Fläche von 2 cm² eine untere Nachweisgrenze von 0,5 mBq erreicht.

Zur Bestimmung von Uranium wird als Alternativmethode die durch UV-Licht angeregte Fluoreszenzstrahlung in Uranium-Verbindungen verwendet. Mit dieser Methode wird eine untere Nachweisgrenze von 1 ng erzielt [20].

*4.3.2. Bestimmung von β-strahlenden Nukliden.* Die Niederaktivitätsbestimmung von $^{89}$Sr-, $^{90}$Sr-, $^{137}$Cs- und $^{210}$Bi-Meßpräparaten mit hoher β-Radionuklidreinheit erfolgt vorwiegend mit integralen Meßmethoden. Dazu werden β-Antikoinzidenzmeßanordnungen, bestehend aus Gasdurchflußmeß- und Gasdurchflußabschirmzählrohr, eingesetzt. Auf Grund der hohen Effektivität der Zählanordnungen und der extrem geringen Untergrundzählrate wird bei 100 min Meßzeit eine untere Nachweisgrenze von 10 mBq erreicht.

Die Bestimmung des Gehaltes von β-strahlenden Nukliden mit einer geringen β-Maximalenergie wie $^{3}$H und $^{14}$C sowie die gleichzeitige Messung eines bekannten Radionuklidgemisches mit β-Maximalenergiedifferenzen $>$ 100 keV erfolgt mit Hilfe der Flüssigkeitsszintillationsspektrometrie. Die untere Nachweisgrenze liegt bei einer Meßzeit von 100 min für $^{3}$H bei 40 mBq.

*4.3.3. Bestimmung von γ-strahlenden Nukliden.* Ein großer Teil der bei der Umweltüberwachung auftretenden Radionuklide kann mit γ-spektrometrischen Meßmethoden bestimmt werden. Den wesentlichen Vorteilen der γ-spektrometrischen Aktivitätsbestimmung wie

- Bestimmung mehrerer gleichzeitig vorliegender Radionuklide und
- Ausschaltung aufwendiger Anreicherungs-, Trenn- und Präparationsverfahren

steht der Nachteil einer im Vergleich zu β-Aktivitätsmessungen ungünstigeren unteren Nachweisgrenze gegenüber.

Enthalten die Proben Radionuklide, deren γ-Energien so weit auseinander liegen, daß keine hohen Anforderungen an das energetische Auflösungsvermögen gestellt werden, werden mit hoher Nachweiseffektivität NaJ(Tl)-Szintillationsdetektoren in starken Pb/Cu-Strahlungsabschirmungen eingesetzt:

- NaJ(Tl)-Bohrlochdetektor (∅ 75 mm × 75 mm) mit Probenwechsler zur Analyse von Pflanzenaschen und Falloutproben mit einem Probenvolumen $\leq$ 5 ml und einer unteren Nachweisgrenze von 10 mBq/g bei 1000 min Meßzeit,
- NaJ(Tl)-Bohrlochdetektor (∅ 175 mm × 150 mm) zur Analyse von Boden-, Gesteins- und Wasserproben mit einem Volumen $\leq$ 250 ml und einer unteren Nachweisgrenze von 0,5 mBq/g bei 1000 min Meßzeit,
- NaJ(Tl)-Detektor (∅ 175 mm × 150 mm) mit einem Ringbechervolumen von 20 l für Wasserproben und einer unteren Nachweisgrenze von 0,05 mBq/g bei 1000 min Meßzeit.

Proben, die ein Radionuklidgemisch enthalten, werden mit hoher Energieauflösung mit Hilfe von GeLi-Halbleiterspektrometern analysiert. Auf Grund der geringeren Detektoreffektivität werden besonders bei Proben mit größeren Volumina schlechtere Nachweisgrenzen im Vergleich zu NaJ(Tl)-Detektoren erreicht. Unter Verwendung starker Abschirmungen (20 cm Pb + 1 cm Cu) zur Nulleffektreduzierung wird bei einer Meßzeit von 1000 min in kleinvolumigen Proben ($\leq$ 5 cm³) eine $^{137}$Cs-Aktivitätskonzentration von 2 mBq/g nachgewiesen.

Die Reduzierung des Strahlungsuntergrundes und damit die Verbesserung der unteren Nachweisgrenze ist einerseits durch Verwendung starker Strahlungsabschirmungen möglich. Andererseits kann die Nachweisgrenze durch elektronische Maßnahmen verbessert werden. Zu diesem Zweck kommt ein Anticomptonspektrometer, bestehend aus einem GeLi-Halbleiterdetektor und einem speziellen, den Halbleiterdetektor umgebenden NaJ(Tl)-Bohrlochdetektor, zum Einsatz. Mit Hilfe einer Antikoinzidenzschaltung wird der Comptonanteil im γ-Spektrum des Halbleiterdetektors um den Faktor 2–3 reduziert und die Nachweisgrenze im Vergleich zu Einfachdetektoranordnungen verbessert.

*4.3.4. Bestimmung von stabilen Elementen.* Zur Bestimmung einer großen Zahl von Bezugselementen und stabilen Elementen, die zur Bewertung ökologischer Belastungswege von Bedeutung sind, sowie zur Ausbeutebestimmung chemischer Trennverfahren wird die Atomabsorptionspektrometrie eingesetzt. Die Konzentrationsbestimmung erfolgt z.B. für solche Elemente wie Ca und K sowie Cs, Sr und Co, deren Konzentrationen im mg/g- bzw. µg/g-Bereich liegen. In Abhängigkeit von den zu bestimmenden stabilen Elementen, dem Probenmedium und der erforderlichen Nachweisgrenze erfolgt die Atomisierung des Probenmaterials in der Flamme ($\geq$ 1 µg/ml) oder in der Graphitrohrküvette ($\geq$ 0,01 µg/ml) [21, 22].

## 5. Bestimmung von Strahlenbelastungen der Bevölkerung

Ausgehend von den in den Überwachungsproben gemessenen Radionuklidkonzentrationen lassen sich einzelne Komponenten der Strahlenbelastung berechnen bzw. abschätzen. Die für die Berechnung mittels Belastungsmodellen erforderlichen Parameter, die von

- meteorologischen Parametern,
- Fallout- bzw. Washoutkonstanten,
- Übergangsfaktoren,
- Atem- bzw. Konsumtionsraten bis zu
- Dosisfaktoren für die Submersion, Bodenstrahlung, Inhalation und Ingestion

reichen, werden in vorbereitenden Untersuchungen ermittelt oder können, falls derartige Faktoren für vergleichbare Bedingungen ermittelt und veröffentlicht wurden, aus der Literatur entnommen werden.

Abb. 5 zeigt in allgemeiner Form das Berechnungsschema. Die Berechnung der äußeren Strahlenbelastung aus der Kontamination der bodennahen Atmosphäre bzw. aus der Falloutkontamination erfolgt im wesentlichen durch Multiplikation der gemessenen Kontamination mit den Submersionsdosisfaktoren oder der Bodenstrahlungskonstante unter Berücksichtigung der Aufenthaltszeit. Auch die Berechnung der inneren Strahlenbelastung infolge der Kontamination der bodennahen Atmosphäre erfolgt durch Multiplikation des gemessenen Kontaminationswertes mit der Atemrate und dem Inhalationsdosisfaktor. Aufwendiger ist die Berechnung der inneren Strahlenbelastung durch Ingestion aus der Flächenkontamination, da bei der Betrachtung der Belastungswege eine weitaus größere Anzahl von Parametern wie

- physikalische und biologische Halbwertszeit,
- Rückhaltefaktor auf der Pflanze,
- Anteil der auf den Boden gelangenden Aktivität,

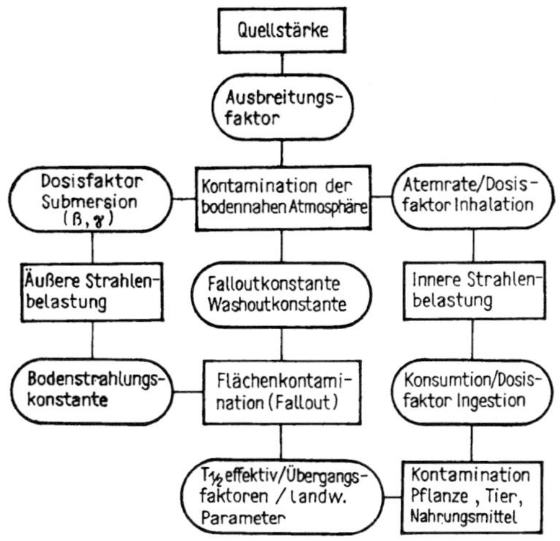


Abb. 5. Schema eines Modells zur Berechnung der Strahlenbelastung

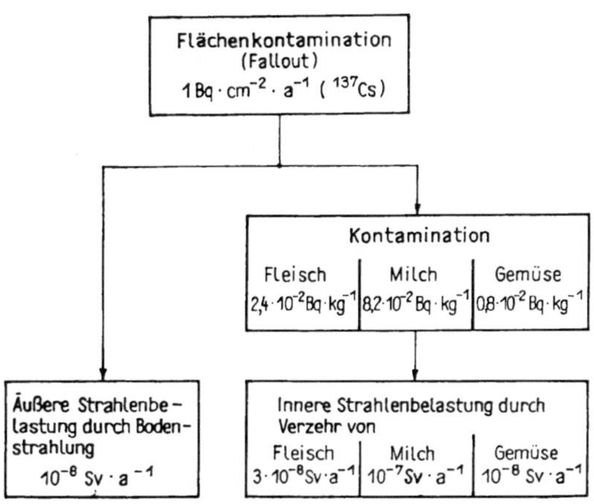


Abb. 6. Innere und äußere Strahlenbelastung durch $^{137}$Cs-Fallout (normiert auf eine Flächenkontamination von 1 Bq · cm$^{-2}$ · a$^{-1}$)

- Vegetationszeit und landwirtschafticher Ertrag,
- Zeit zwischen Ernte und Verzehr sowie
- Übergangsfaktoren Boden/Pflanze und Pflanze/Tier

zu berücksichtigen sind.

In Abb. 6 sind als Beispiel die Berechnungsergebnisse der äußeren Strahlenbelastung und der inneren Strahlenbelastung durch Ingestion aus der Flächenkontamination für das aus dem globalen Fallout kommende Radionuklid $^{137}$Cs dargestellt. Es wird deutlich, daß in diesem Fall die innere Strahlenbelastung durch Ingestion wesentlich größer als die äußere Strahlenbelastung ist und folglich die Falloutmessung eine notwendige Komponente der Überwachung sein muß.

In Verallgemeinerung dieser Berechnung auf alle Quellen der zivilisationsbedingten Strahlenbelastung und unter Verwendung der Ergebnisse von probenanalytischen Radionuklidbestimmungen und Literaturwerten [23] ist in Tab. 3 abschließend die mittlere jährliche effektive Strahlenbelastung von Personen der Bevölkerung der DDR angegeben. Sie betrug 1984 ca. 2,5 mSv, wobei sich

28% aus der natürlichen Strahlenbelastung,
34% aus der zivilisationsbedingten Strahlenbelastung aus natürlichen Quellen und
38% aus der zivilisationsbedingten Strahlenbelastung aus künstlichen Quellen

ergeben.

Eingegangen am 9. 5. 1985
Überarbeitete Fassung am 10. 7. 1985

## Literatur

[1] Verordnung über die Gewährleistung von Atomsicherheit und Strahlenschutz vom 11. Oktober 1984. GBl. I, Nr. 30, S. 341
[2] Objektives and Design of Environmental Monitoring Programms for Radioactive Contaminants. SAFETY SERIES No. 41. Vienna, IAEA 1975
[3] Richtlinie zur Umgebungsüberwachung von Kernanlagen vom 8. 7. 1974. Mitteilungen des Staatlichen Amtes für Atomsicherheit und Strahlenschutz der DDR, Nr. 7, 1975
[4] *W. Rehak, F. Moldenhawer*, SZS-157 (1974) 258
[5] *D. Herrmann, D. Obrikat, W. Rehak*, SZS-157 (1974) 264
[6] *W. Blume, D. Obrikat*, Isotopenpraxis **15** (1978) 374
[7] *F. Dittel*, Fresenio Z. analyt. Chem. **228** (1967) 432
[8] *C. Gleit, W. Holland*, Analyt. Chem. **34** (1962) 1454
[9] *G. G. Rocco, W. S. Broecker*, J. Geophys. Res. **68** (1963) No. 15, 4501
[10] *J. Kollar, M. Tatara, D. Chorvat, T. Trnovec, M. Szlaurova*, SZS-157 (1974) 232
[11] *R. L. Blanchard*, Public Health Serivce Publication No. 999-RH-9 (1964) 19
[12] *W. Rehak*, SZS, Nr. 1 (1967)
[13] *M. Cospito, L. Rigali*, Anal. Chim. Acta **106** (1979) 385
[14] *J. C. Veselsky, P. C. Kirl, N. Sezginer*, J. Radioanal. Chem. **21** (1974) 97
[15] *W. Rehak, B. Feddersen*, SZS Nr. 1 (1970)
[16] *W. Rehak, B. Feddersen*, SZS Nr. 3 (1972)
[17] *W. Blume*, SZS Nr. 2 (1966)
[18] *F. B. Johns, P. B. Hahn* et al., EMSL-LV-0539-17 (1979)
[19] Methoden zur Bestimmung von $^{90}$Sr, $^{137}$Cs, $^{226}$Ra und U in Umgebungsproben. SZS (1972)
[20] *H. Scheel*, SZS Nr. 4 (1964)
[21] *H. Gegusch*, Radioökologische und chemisch-analytische Untersuchungen an ausgewählten Materialien. Diss. Humboldt-Universität Berlin, 1979
[22] *H. Gegusch, B. Findeisen*, Comparison of Different Graphite Materials for Tube and Platform in the Graphite Furnace for the Determination of Cobalt, Nickel and Iron by AAS. VIIIth CANAS, Ceske Budyovice, 18.–22. 06. 1984
[23] *J. Steuer, W. Angerstein*, Radiologica Diagnostica **25** (1984) 475
[24] United Nations Scientific Committee on the Effects of Atomic Radiation, 1982 Report. United Nations Publication, Sales No. E.82.IX.8. 06300 P.

Tab. 3. Mittlere Strahlenbelastung von Personen der Bevölkerung der DDR

| Belastungsart | | Quellen | mittlere jährliche effektive Äquivalentdosis, µSv | Bemerkungen |
|---|---|---|---|---|
| natürliche Strahlenbelastung | | Kosmische Strahlung | 240 | Erhebungsmessungen und Berechnungen des SAAS |
| | | terrestrische Strahlung | 60 | |
| | | Ingestion (nat. Radionuklide) | 370 | |
| | | Inhalation (nat. Radionuklide) im Freien | 80 | |
| zivilisationsbedingte Strahlenbelastung | natürliche Quellen | Aufenthalt in Häusern | 890 | Ergebnisse von Erhebungsmessungen des SAAS |
| | | fossile Brennstoffe | < 2 | |
| | | Phosphatdünger | < 1 | |
| | künstliche Quellen | globaler Fallout | < 5 | Überwachungsergebnisse des SAAS |
| | | Kernanlagen | < 1 | |
| | | Medizinische Diagnostik | ~ 1000 | [23] |
| | | andere Quellen | ~ 10 | [24] |

Isotopenpraxis 22 (1986) 3, S. 106 bis 107

# Tagungsberichte

## Jahrestagung der Gesellschaft für Ernährung der DDR in Leipzig vom 8.—11. September 1985

Der mit internationaler Beteiligung an der Karl-Marx-Universität Leipzig durchgeführte Kongreß stand unter dem Leitthema „Eiweißversorgung bei Tier und Mensch“. In einer Plenarsitzung zu Beginn der Tagung wurden Probleme und medizinische Aspekte der Eiweißversorgung des Menschen sowie neue Erkenntnisse zur Verdauung und Resorption behandelt. Darüber hinaus wurde auf die Eiweißbedarfsdeckung der Tierbestände der DDR eingegangen.

Die Arbeit der Sektionen Human- und Tierernährung wurde in einer gemeinsamen Sitzung insbesondere durch die Themen zur Bewertung von Eiweißressourcen und der Eiweißqualität für die tierische und menschliche Ernährung, die Nutzung neuer Nahrungsmittel und zur Beziehung zwischen Energie- und Eiweißumsatz bestimmt. Auch standen Fragen des Einsatzes tracertechnischer Methoden zur Bestimmung des Proteinturnovers, einschließlich der mathematischen Modellierung des Ganzkörperproteinstoffwechsels im Mittelpunkt des Interesses. Vorträge zu speziellen Problemen der Verdauung und Resorption von Aminosäuren und Proteinen ergänzten das umfangreiche Programm.

In der Sektion Krankenernährung standen der Eiweißbedarf und die Eiweißsubstitution beim Diabetes mellitus, bei gastroenterologischen und Lebererkrankungen sowie Möglichkeiten der prä- und postoperativen Eiweißversorgung des Patienten als Schwerpunkte zur Diskussion.

In den Sitzungen der Arbeitsgemeinschaft „Parenterale und Sondenernährung“ wurden diagnostische Maßnahmen der enteralen und parenteralen Nahrungszufuhr bei verschiedenen gastroeneterologischen Erkrankungen und bei Niereninsuffizienz beraten. Dabei kam dem Problem der Spurenelementdiagnostik im Zusammenhang mit der Bewertung der parenteralen Ernährung in der klinischen Praxis eine besondere Bedeutung zu. Weitere Themenkomplexe waren dem Postaggressions- und Fettstoffwechsel aus ernährungsphysiologischer Sicht gewidmet.

In der Sektion Ernährungswirtschaft wurden ernährungsökonomische Aspekte der Eiweißversorgung diskutiert.

Das umfangreiche Programm der Sektion Tierernährung konzentrierte sich auf Beiträge zur Verdauung und Resorption von Eiweiß und Aminosäuren bei Monogastriden und Wiederkäuern einschließlich der Bewertung von Eiweißfuttermitteln und des Eiweiß- und Energieumsatzes.

Aus isotopentechnischer Sicht kommt dem stabilen Stickstoffisotop $^{15}N$ zur Lösung der Eiweißproblematik bei Tier und Mensch eine herausragende Stellung zu. Insbesondere scheint sich für den $^{15}N$-Einsatz zur Bestimmung des nutritiven Effektes der parenteralen und enteralen Nahrungszufuhr in der Kranken- und Säuglingsernährung, ausgehend von unterschiedlich markierten Proteinstoffwechselprecursoren, ein neues, für die klinische Praxis bedeutsames Anwendungsgebiet der Isotopentechnik zu eröffnen.

Das umfangreiche Programm wurde durch mehr als 40 Posterbeiträge zu allen behandelten Themen und zu methodischen Fragen ergänzt.

*K. Jung*

Eingegangen am 24. 9. 1985

## II. Hepatologisches Symposium, 3.—5. September 1985 in Jena

Auf Einladung der Biochemischen Gesellschaft der DDR und der Gesellschaft für Klinische Chemie und Laboratoriumsdiagnostik der DDR trafen sich zum zweiten Male innerhalb von 3 Jahren die Spezialisten der Leberforschung (160 DDR-Teilnehmer sowie geladene ausländische Gäste).

Der Vorteil solcher Symposien zu einem eng begrenzten Fachgebiet liegt auf der Hand: Auf der Grundlage des gegenseitigen wissenschaftlichen Verständnisses und Interesses wird zu nahezu allen Beiträgen (42 Vorträge und 50 Poster) intensiv diskutiert und der Wissensstand aktiv vorangebracht. Die Biochemiker und anderen Naturwissenschaftler bieten neue diagnostische Verfahren an, während die Kliniker mit Erfahrungsberichten nachweisen, daß noch immer nicht die „idealen“ Parameter zur Charakterisierung der unterschiedlichsten Leberfunktionen bzw. pathologischen Zustände zur Diagnostik bereitstehen. Auch die Pathologen demonstrierten bei allem Erkenntniszuwachs ihre Grenzen.

Von den ausländischen Beiträgen seien die von *B. B. Dineva* (Sofia, VRB), *D. Häussinger* (Freiburg, BRD) und *A. Hesz* (Kecskemét, UVR) zum Proteinstoffwechsel sowie von *L. Milwan* (Moskau, UdSSR), *P. D. Duee* (Meudon-Bellevue, Frankreich), *M. Hohenegger* (Wien, Österreich) und *E. Pawlicka* (Bialystok, VRP) zum Energie- und Kohlenhydratstoffwechsel hervorgehoben.

Kaum eine der vorgestellten Untersuchungen kam ohne einen „Marker“ aus. Darunter werden allerdings nicht nur isotope Markierungen verstanden, sondern z. B. auch Konjugatbildungen von Lipoproteinen mit Goldkolloid zur Sichtbarmachung oder die Enzymimmunoassays. In 22 Beiträgen wurden Isotopenmethoden diskutiert — das ist immerhin nahezu ein Viertel aller Beiträge. Noch vor $^{3}H$, $^{14}C$, $^{125}I$ und $^{131}I$ war mit 7 Arbeiten das stabile Stickstoffisotop $^{15}N$ am häufigsten vertreten. So wurden vom Zentralinstitut für Isotopen- und Strahlenforschung Leipzig der AdW (*H. Faust* u. Mitarb.) in Kooperation mit Leipziger Kliniken drei neuartige Leberfunktionstests vorgestellt ([$^{15}N$]Methacetin-Test, [$^{15}N$]Ammonium-Test und [$^{15}N$]Hippursäure-Test). 5 Arbeiten wiesen die Nutzung von $^{14}C$ zur Klärung von Detailfragen des Kohlenhydrat- und Fettstoffwechsels aus.

Insgesamt zeigt dieses Symposium ebenso wie die Tagung „Eiweißversorgung von Mensch und Tier“ (siehe den vorhergehenden Bericht) wiederum, daß sich die Isotopenpraxis (im wörtlichen Sinne!) zunehmend auch in den Tagungen und Veröffentlichungen derjenigen Fachrichtungen widerspiegelt, für die in der Vergangenheit zielgerichtet Isotopenmethoden entwickelt wurden und — das haben beispielsweise zum Hepatologischen Symposium die Reaktionen der Kliniker gezeigt — auch weiterhin entwickelt werden müssen.

*P. Krumbiegel*

Eingegangen am 25. 9. 1985

Isotopenpraxis 22 (1986) 3, S. 107 bis 108

# Buchbesprechungen

## Detektoren für Teilchenstrahlung

*K. Kleinknecht. Stuttgart: B. G. Teubner 1984; 216 S., 137 Abb.; 16 Tab.; kart. DM 26,80*

Der Buchinhalt ist wie folgt eingeteilt: 1. Grundlagen, 2. Ionisationsmessung, 3. Ortsmessung, 4. Zeitmessung, 5. Teilchenidentifizierung, 6. Energiemessung, 7. Impulsmessung und 8. Beispiele für Anwendungen von Detektorsystemen. Er beruht auf Vorlesungen, die der Autor seit rund zehn Jahren an der Dortmunder Universität hält. Der Text bevorzugt die Belange der Physik an Beschleunigern und ist insoweit nicht repräsentativ für die Kernstrahlungsdetektoren und ihre Anwendungen schlechthin. Man erfährt aus dem Buche leider nichts über die Existenz von Gasszintillatoren, Proportionalszintillationsdetektoren, Festkörperspurdetektoren und die Teilchenregistrierung mittels Photoemulsionen, geschweige über deren Eigenschaften oder Anwendungen, obwohl z.B. die Photoplatte historisch nicht nur den ältesten Detektor darstellt, sondern in weiteren Bereichen der Autoradiographie und der Hochenergiephysik noch immer der Detektor der Wahl ist. Ihr Übergehen kann insoweit nichts mit Modernität zu tun haben. Die Überschriften des Buches und seiner Kapitel sind teilweise verblüffend, wenn nicht irreführend. Warum ist die Rede von Teilchendetektoren, wenn Photonen als „ungeladene Teilchen" am Ende noch mit einbezogen sind? Und wer vermutet, daß sich hinter der Überschrift ‚Zeitmessung' bis auf ein paar Bemerkungen zu Funkenzählern nichts anderes als Ausführungen über den Szintillationszähler verbergen? Nun könnte man ja auf den Gedanken kommen, daß sich das Einteilungsprinzip vielleicht an den Anwendungen in der Beschleuniger-Physik orientiert, aber weit gefehlt: Wechselwirkungseffekte (z.B. Ionisation) und Anwendungsgesichtspunkte sind bunt gemischt.

Das Beste am Text ist nach Meinung des Rezensenten die einheitliche Darstellung der Wechselwirkungseffekte geladener Teilchen mit Materie auf der Grundlage der dielektrischen Stoffeigenschaften, eine Darstellungsweise, die auf vom Autor zitierten Arbeiten von *Allison, Cobb* und *Wright* (Ann. Rev. Nucl. Sci. *30* (1980) 253 und Oxford-Preprint OUNP 35/83 (1983)) beruht. Leider bleiben Wechselwirkungen mit Kernen, die im gegebenen Zusammenhang mindestens für das Bremsvermögen schwerer Teilchen geringerer Energie wesentlich sind, unerwähnt.

Das explizite Anwendungskapitel am Schluß des Buches hat locker-beschreibenden Charakter: Mit wenig Bezug zum vorherigen Text werden Einzelfälle qualitativ und kurz behandelt.

Das Sachverzeichnis von einer knappen Seite entspricht im hier dargestellten Sinne der Gesamtanlage: Auslösezählrohr, Proportionalzählrohr oder nur Zähler allgemein finden sich darin nicht, dafür aber z.B. das Stichwort Stereowinkel.

Für eine allgemein brauchbare Ausbildung von Studenten einschlägiger physikalischer Richtungen möchte man sich ein wenig mehr Systematik und Vollständigkeit selbst dann wünschen, wenn zufolge anderer und thematisch angrenzender Vorlesungen vielleicht hier und da schon einiges über Detektoren vorweggenommen sein sollte. Das gilt natürlich um so mehr für den potentiellen Leserkreis der Nicht-Dortmunder Physikstudenten. *G. Brunner*

## Zur Metallogenie der Sn-W-Mineralisationen auf dem Territorium der MVR

*D. Sengee. Leipzig: VEB Deutscher Verlag für Grundstoffindustrie 1984. 77 S., 21 Abb., 4 Tab., 36 Lit. (Freiberger Forschungshefte C 394); kart. M 24,—*

Der aus Ulan-Bator (MVR = Mongolische Volksrepublik) stammende Autor legt einen Auszug seiner 1982 an der Bergakademie Freiberg verteidigten Dissertation vor. Auf den ersten Seiten erhält der Leser einen geologisch-tektonischen Überblick über das Territorium der MVR. Eine Übersicht über die metallogenetische Rayonierung der MVR schließt sich an. Im folgenden wird nur der nördliche und östliche Teil des Landes besprochen, der zur Mongolo-Transbaikalischen metallogenetischen Provinz gehört. Hier entstanden im Mesozoikum drei im Durchmesser jeweils 300 bis 600 km messende Aufwölbungszonen mit granitoiden Intrusionen. Der Hauptteil der Arbeit behandelt ein Kristallinmassiv in der mittleren dieser Beulen, der Chêntêj-Aufwölbung. Das Žančivlan-Massiv (500 km$^2$), zum frühmesozoischen Intrusionszyklus gehörig (Trias bis unt. Jura), ist durch eine dreiphasige granitoide Intrusion gekennzeichnet, deren erste mit nur schwach vererzten Pegmatiten, die zweite mit Sn-W-Vererzungen führenden Greisen und pneumatolytisch-hydrothermalen Quarzgängen und die dritte mit einer Albitisierung und Nb-Ta-Sn-Mineralisation verknüpft ist.

K-Ar-Altersbestimmungen ergaben für die Granitoide ein Alter von 228—188 Mill. Jahren, übereinstimmend ermittelt an Biotit, Kalifeldspat und Muskovit. Die drei Intrusionsphasen lassen sich hierbei nicht unterscheiden. Auch die Erzbildung wurde datiert: Die Greisenbildung ergab ein Alter von 205—196 Mill. Jahren; eine zweite Phase der Erzbildung ergab 164—126 Mill. Jahre (bis Oberkreide). Spätwirkungen kommen in Verquarzungen der hangenden Braunkohle zum Ausdruck.

Die Arbeit enthält eine petrographisch-geochemische Charakterisierung der verschiedenen Granitoide (jeweils 15 Elemente werden betrachtet) und eine mineralogischlagerstättenkundliche Beschreibung des Žančivlan-Gebiets, wobei sieben Mineralisationstypen mit ihren Bildungsabfolgen und Mineralgehalten unterschieden werden. Ein Abschnitt über erzkontrollierende Faktoren und Erzindikatoren beschließt das Heft. Hier werden die von *Tischendorf*, *Baumann* und anderen für das Erzgebirge entwickelten metallogenetischen Kriterien übertragen und nutzbar gemacht.

Die Arbeit berücksichtigt besonders die sowjetische Literatur (methodisch und regional) und leistet einen wertvollen Beitrag zur Gliederung eines mehrphasig und komplex vererzten Gebietes. Das Heft ist für Petrologen und Geochemiker, die sich über saure Kristallinkomplexe informieren wollen, sowie für Vertreter der Mineralogie und Erzlagerstättenkunde zu empfehlen. *A. Müller*

### Einführung in die Kristallographie

*W. Kleber. 16., durchgesehene Auflage. Berlin: VEB Verlag Technik 1985. 384 Seiten, 375 Abb., 39 Tafeln, 3 Beilagen; Kunstleder M 29,—*

Das aus der Mineralogie hervorgegangene Fachgebiet Kristallographie hat sich in den letzten Jahrzehnten zu einer wichtigen selbständigen Wissenschaftsdisziplin entwickelt. Zwischen ihr und inzwischen einer großen Zahl anderer naturwissenschaftlicher und technischer Fachgebiete haben sich enge, wechselseitige Beziehungen herausgebildet, ohne die viele neue Entwicklungen, vor allem im Bereich der Werkstoffe, aber auch in technischen Zweigen wie z. B. der Aufbereitung, der Baustoff- und Keramikindustrie, der chemischen Verfahrenstechnik, der optischen Industrie bis hin zur Mikroelektronik und dem wissenschaftlichen Gerätebau, nicht möglich gewesen wären. Speziell für den Kerntechniker spielen bei der Beurteilung des Werkstoffverhaltens in Strahlungsfeldern oder der Elementverteilung bei Erstarrungsvorgängen — zwei aus einer Vielzahl kern- und isotopentechnischer Probleme herausgegriffene Beispiele — Strukturfragen eine z. T. dominierende Rolle.

Dem interdisziplinären Charakter der Kristallographie wurde bei der Anlage des Buches weitgehend Rechnung getragen. Im Kapitel 1 ,,Kristallstrukturlehre und Kristallmorphologie“ wird der Gitterbau der Kristalle behandelt und in Kapitel 2 ,,Kristallchemie“ dieses Teilgebiet aus dem Prinzip der Kugelpackungen und den Bindungszuständen entwickelt. Das 3. Kapitel ,,Physikalisch-chemische Kristallographie“ ist der Realstruktur von Kristallen sowie der Kristallisation und Kristallzüchtung gewidmet. Im Kapitel 4 wird an Hand der thermischen Ausdehnung und der Wärmeleitung der Tensorformalismus anwendungsbezogen eingeführt. Die Abschnitte über die elektrischen und magnetischen Eigenschaften von Kristallen tragen neueren Entwicklungen auf diesen Gebieten Rechnung. Einem Abschnitt über Kristalloptik folgt die Behandlung der stimulierten Lichtemission (Laser) und der Photochromie. Das abschließende Kapitel 5 ,,Strukturanalyse von Kristallen“ befaßt sich mit der Röntgenkristallstrukturanalyse, der röntgenographischen Untersuchung der Realstruktur und der Untersuchung von Kristallen mit Korpuskularstrahlen. Ein Literaturverzeichnis über weiterführende und ergänzende sowie zitierte Publikationen und ein Sachwörterverzeichnis runden das Werk ab.

Das Buch ist vom Autor bei Bemühung um Verständlichkeit in didaktisch vorbildlichem Stil verfaßt. Es wird als Nachschlage- und Lehrbuch Interessenten aus der Kristallographie selbst, der Mineralogie, Chemie, Physik und Metallkunde sowie den technischen Disziplinen Bergbau, Aufbereitung, Silikattechnik, Technische Keramik, Metallurgie, Elektrotechnik, Maschinen- und Gerätebau sehr empfohlen. *J. Holzhey*

### Physikalische Methoden in der Chemie

*Hrsg.: B. Schröder u. J. Rudolph. Weinheim — Deerfield Beach, Florida — Basel: VCH Verlagsgesellschaft/Physik Verl. 1985. IX, 322 S., 240 Abb., 8 Tab.; geb., DM 48,—*

Noch vor wenigen Jahrzehnten war es völlig ausreichend, wenn ein Organiker den Schmelzpunkt einer festen Verbindung, den Brechungsindex einer Flüssigkeit und möglicherweise ein UV-Spektrum zur näheren Charakterisierung seines Stoffes angab. Dies hat sich grundlegend verändert; heute werden dem Chemiker eine Fülle physikalischer Methoden zur Anwendung angeboten, die zu überschauen und mit Sachkenntnis zu interpretieren vertiefte Kenntnisse erfordern.

Der Verlag Chemie gibt seit 1967 u. a. die Zeitschrift ,,Chemie in unserer Zeit“ heraus, in der eine Artikelserie den physikalischen Methoden in der Chemie gewidmet ist. Nicht jedem ist diese Zeitschrift zugänglich, aber die ausgewählten Beiträge über diesen Themenkreis haben einen hohen didaktischen und zeitlosen Wert. So haben sich die Herausgeber dieses Sammelbandes der aktualisierten Zusammenstellung der in der genannten Zeitschrift erschienenen Beiträge unterzogen, und es liegt ein gutes, leicht verständliches und reich bebildertes Buch vor. IR- und *Raman*-Spektroskopie, Kernresonanz und $^{13}$C-NMR-Spektroskopie, die Massenspektrometrie, EPR-Spektroskopie, ESCA- und UV-Photoelektronenspektroskopie, die chisoptischen Methoden in der Stereochemie, *Mößbauer*-Spektroskopie und schließlich die Röntgenstrukturanalyse von Molekülen werden knapp, aber leicht erfaßbar behandelt. Ausgewählte und weiterführende Literatur wird dem Lernenden durch eine weiterreichende Vertiefung in die einzelnen Arbeitsmethoden angeboten — insgesamt 148 Zitate, die bis in die Jahre 1981 und 1983 reichen; das Hauptfeld liegt aber in den 70er Jahren, da in diesem Dezennium der größte Teil der Beiträge in der genannten Zeitschrift erschienen ist.

Wenn auch die Eskalation der physikalischen Methoden in der Chemie weitergeht und diese sehr schnell zu Routinemethoden avancieren: PIXE, SIMS, ENDOR-Spektroskopie, diffuse Kleinwinkelstreuung von Makromolekülen in Lösungen, photoakustische Spektroskopie etc., um nur einige zu nennen: hier ist für den Studierenden und auch für den Lernenden ein Grundstock gelegt, der leicht — aufbauend darauf — weitergeführt werden kann und auch sollte. Eine schöne Zusammenstellung, die ihre begeisterte und zustimmende Leserschaft finden wird; denn es ist erstaunlich, wie mit leichter Hand auch schwierige Zusammenhänge dargestellt werden können. *H. Hübner*

# ISOTOPENPRAXIS

Band 22 | 1986 | Heft 3


## Contents / Inhalt




**Isotopenpraxis is indexed in Current Contents/PC & ES, Science Citation Index, and INIS Atomindex**

FSC
www.fsc.org
MIX
Papier | Fördert gute Waldnutzung
FSC® C083411

Zeitfracht Medien GmbH
Ferdinand-Jühlke-Straße 7
99095 Erfurt, Deutschland
produktsicherheit@kolibri360.de